Fire Department Pumping Apparatus Maintenance

First Edition

By
Don Henry

Published by
Fire Protection Publications • Oklahoma State University •
Stillwater, Oklahoma

Project Manager: Cindy Pickering
Development Editor: John Joerschke
fire etc. Editor: Natalie Clennett
Design/Layout: Ann Moffat
Graphics: Ann Moffat, Brien McDowell

ISBN 0-07939-219-3
Library of Congress Control Number: 2003109747
First Edition
First Printing, October 2003

10   9   8   7   6   5   4   3   2   1   *Printed in the United States of America*

Contents

About the Author

Don Henry has taught automotive service technician and heavy equipment technician at Lakeland College for 17 years. Canada's only college-level fire truck maintenance program was codeveloped and delivered by Don. He has been a member of the Society of Tribologists and Lubrication Engineers for 12 years. He has been a member of the Fluid Power Society for 13 years and holds a certified fluid power specialist exam rating. Don holds an emergency vehicle technician rating as a master technician. Don is a principal member of the National Fire Protection Association 1071 on Emergency Vehicle Mechanic Qualification for North America, the past president of the National Association of Emergency Vehicle Technicians, and the past chairman of the Apparatus Maintenance Section for the International Association of Fire Chiefs. Don holds journeyman certification in both automotive service technician and heavy equipment technician trades.

Acknowledgements

The author wishes to acknowledge the following people and organizations for their technical assistance, advice, direction, and encouragement:

Chris Atkin
Major, OMM, CD
Loyal Edmonton Regiment
Edmonton, Alberta

Jim Parker
Vice President
Pierce Manufacturing

Greg Pfaff
Pierce Manufacturing

Bob Linster
Hale Products Company

Lairy Normand
W.S. Darley and Company

Gordon A. Hills
Lt. Col. (ret), CD, Ph.D., P.Ag.
Lakeland College
Vermilion, Alberta

Paul Myshaniuk
Instructor
Auto/Diesel Dept.
Lakeland College
Vermilion, Alberta

Ken Lawrence
Instructor
Fire Apparatus Maintenance
Lakeland College
Vermilion, Alberta

Don Gnatiuk
C.E.O. fire etc.
(Alberta Fire Training School)
Vermilion, Alberta

Gary A Hovdebo
Owner
Hill's Hot Shot Service
Sherwood Park, Alberta

Leroy Carlson
Chief of Services
Edmonton Emergency Response Department
Edmonton, Alberta

Boyd Cole
Chairman
NFPA 1071 Committee

Dan Juntune
Waterous Pump Company

Gregg Geske
Waterous Pump Company

Nate Berry
Fire Research Corporation

Gary Ewers
Class 1, Inc.

Andy Holzli
Superior Emergency Vehicles
Red Deer, Alberta

Warren Clement
Pumped and Wired Emergency Vehicles
Nova Scotia

CTC Analytical Services, Inc.
Canada

AGAT Laboratories
Edmonton, Alberta

Patrick Kilbane
Predict
Cleveland, Ohio

Dan Cromp
Chief Engineer
Braidwood Fire Department

Skee Stanley
Monterey Airport Fire Department

Tom Gaines
Tulsa, Oklahoma

Dave Sweden
Service Manger
CT Fire Service

Brenda K. Smith
Allison Transmission Division
General Motors

Phil Wagner
General Manager
American Lafrance, Michigan

Fluid Life Corporation
Edmonton, Alberta

Bob Scott
Lube Works
Edmonton, Alberta

Jim Darley
W.S. Darley Pump

Bill Barnes
MC Products

Gerald R. Lee
California

Dane Jones
Chairman
Apparatus Maintenance Section
International Association of Fire Chiefs

Fluid Power Society
Cherry Hills, New Jersey

Society of Tribologists and Lubrication Engineers
Alberta Section

National Association of Emergency Vehicle Technicians

Steve Wilde
Emergency Vehicle Technicians Certification Commission

Terry Munro
ULC Canada
Calgary, Alberta

Gary Handwerk
Director Application Engineering
Hale Products, Inc.

Dennis Litchenstine
C.E. Neihoff Company

Mark Smith
Analysts
Torrance, California

Leonard Regier
Waterous Detroit Diesel
Calgary, Alberta

Further editorial assistance was provided by the following, fire etc. (formerly, Alberta Fire Training School) Staff:

Chris Senaratne
Manager
Accreditation and Course Development

Mariette Fisher
Course Developer

Mariette Corbiere
Administrative Support
Course Development

Natalie Clennett
Course Research and Development

Dedication

This book is dedicated to my loving and supportive wife, Blanche, without whose patience and understanding this project never would have come to fruition. I owe her a debt of gratitude for all of the household chores left undone and social events foregone.

On the technical front I must also acknowledge the inspiration given to me by an absolute giant in the fire service, Boyd Cole. Without Boyd's encouragement this project would not have even started.

Chapter 1
General Knowledge and Safety Requirements

General Knowledge and Safety Requirements

Introduction

Fire apparatus are complicated vehicles; therefore, the people who maintain them need to have a unique set of skills and knowledge. An understanding of the history of the development of the standards in place today can guide you in acquiring the skills and knowledge you will need as an emergency vehicles technician. It is also important to understand these standards' role in keeping our communities safe.

As a result of the dedication and perseverance of many different groups whose aims were to help technicians gain needed skills and knowledge, a set of standards has been created. In 1988 the International Association of Fire Chiefs (IAFC) formed the Apparatus Maintenance Section (AMS). The section's mandate is threefold:

1. National certification of fire apparatus mechanics

2. Promulgation of training and education regarding equipment and apparatus

3. Exchange and distribution of information regarding equipment and apparatus

In 1991 the National Safety Transportation Board of the United States of America issued its *Special Investigation Report, Emergency Fire Apparatus.* It urged both the Federal Emergency Management Agency (FEMA) and the IAFC to encourage the maintenance and inspection of fire apparatus. The AMS section responded by creating the Emergency Vehicle Technician Certification Commission (EVTCC). EVTCC subsequently created an excellent set of exams to challenge the technician's knowledge of fire apparatus maintenance and repair. Passing these exams should be the minimum requirement for anyone wishing to work on the maintenance and repair of fire apparatus. This qualification assures the taxpayer and the fire department that the capital investment in equipment and the safety of the community are being protected.

In early 1993 the National Fire Protection Association (NFPA) created two committees to develop standards for emergency vehicle repair and maintenance. The task of the NFPA 1071 Committee was to create a professional qualification standard for the emergency vehicle technician. The scope of this standard NFPA 1071, *Standard for Emergency Vehicle Technician Professional Qualifications,* is to identify the minimum job performance requirements for a person to be qualified as an EVT. The purpose of the standard is to ensure that personnel who are engaged in the inspection, diagnosis, maintenance, repair, and testing of emergency response vehicles are qualified. This standard committee examined the duties and tasks of an EVT and identified the knowledge and skills necessary to accomplish these duties (Appendix A summarizes NFPA 1071).

The task of the NFPA 1915 Committee was to create a standard for fire apparatus maintenance. The scope of the NFPA 1915, *Standard for Fire Apparatus Preventive Maintenance Program,* is to define the minimum requirements for establishing a preventive maintenance program. Its purpose is to ensure that fire apparatus are maintained in a safe operating condition and ready for response at all times. The committee achieved this by identifying the systems and items to be inspected, the frequency of maintenance, and the requirements for testing. In many cases, the NFPA adopted the Federal Motor Vehicle Safety (FMVS) Standards, which has helped to counter some departments' mentality that they are exempt from those standards. The committee's most important work, however, was to set out-of-service (OOS) standards that exceed the FMVS standards in order to meet the needs of fire service vehicles.

Both NFPA 1071 and NFPA 1915 were created by a cross-section of the industry. End users, enforcement agencies, educators, and manufacturers were all involved to ensure these standards were fair and attainable. As with any NFPA standards, these are *minimum* standards. They should be considered not as the level the department should strive to achieve but as the base level below which the department never wants to fall. The NFPA adopted these standards in May 2000 and updates them periodically to ensure that they are meeting the needs of the fire service.

In 1991 the National Association of Emergency Vehicle Technicians (NAEVT) was formed. Its purpose as a national membership network is to obtain and facilitate training for the technician. It maintains an active Web page with an informative bulletin board. This free exchange of ideas is useful to the technician. A question posted on the Web page will receive a response from someone who either has had this problem and solved it or also wants to know the answer. Progressive manufacturers also monitor this board. At times it is possible to sense the technicians' level of frustration as they seek answers to their problems, but this indicates that they care about their equipment's state of readiness and are committed to their communities' safety.

The fire service is somewhat different from commercial industries because in most departments the final say on any matter rests ultimately with the fire chief. This person is usually identified as the authority having jurisdiction. That means that the buck stops at the fire chief's desk. Most fire chiefs are conscientious people who are fully aware of their public trust. Fire chiefs and town councillors should ask themselves the following questions to ensure the department's maintenance of fire apparatus is fulfilling that public trust:

1. Are we satisfied with the present level of maintenance?

2. Are we satisfied with the cost of our present level of maintenance?

3. Do we predict that in the next decade the age of our front line apparatus will decrease?

4. Do I, as the fire chief, have a program, plan, and money to train new technicians?

If they answer no to any of those questions, they need to provide serious training for their technicians. Let's look at each question in detail.

Question 1: Are we satisfied with the present level of maintenance? A record of the times the trucks failed to perform must be kept. This is often called a failure-to-roll rate. An old adage is, "If you cannot measure it, you cannot improve it." Keep the records and find out what the problems are; then solve them.

Question 2: Are we satisfied with the cost of the maintenance? Perhaps the trucks have a very low failure-to-roll rate, but it is just too expensive to drive them back to the repair depot in the next major city to get every little thing fixed. Maybe the department should train its own technicians. Having trained persons on staff would be far better than waiting for a service truck to arrive from miles away. By doing this maintenance locally, the department could free resources to purchase needed equipment for its firefighters.

Question 3: Do we predict that in the next decade the age of our front line apparatus will decrease? The answer to this question reveals a dilemma that many departments face. In most cases they have minimum resources to accomplish maximum results; they are always looking for the biggest bang for the buck. Older equipment will need more maintenance, but new equipment will need trained technicians to repair the modern electronics such as pump controls, multiplexed systems, and the newer compressed air foam systems. Either way, technicians will need to be trained.

Question 4: Do I, as the fire chief, have a program, plan, and money to train new technicians? The answer to this question indicates whether the fire chief can make long-range plans. If the chief's definition of long-range planning is a week from Tuesday, then the department is in trouble. If the shop employs young eager technicians to repair the department's trucks, it indicates that some forward planning has occurred. As the nation's work force ages, it will be necessary to replace older workers with skilled younger people. If a department's culture fosters an atmosphere of training, it will get and keep good technicians. The same tactics used to recruit young firefighters will work here, too: training, promotion, reward, and recognition. Some departments think nothing of spending thousands of dollars and man hours every year to train their firefighters, but when asked how much they spend keeping their technicians up to speed, they reply, "Oh, that's the town's problem." In fact, it's the fire chief's problem. As the authority having jurisdiction's representative, the chief is responsible for apparatus maintenance.

NFPA 1071 was developed to fit large city, small town, and rural departments. All three types of departments can benefit by having trained technicians. For example, a large city department not far from where I live had an excellent full-time fire apparatus maintenance department with twelve competent technicians. Their failure-to-roll rate was not even close to one percent. The city decided to tighten the fiscal belt, however, and an outside consultant recommended combining the fire service mechanics with the technicians in the city's mobile fleet maintenance shop. Because the two groups of technicians had the same qualifications (Canadian Inter-Provincial Red Seal for Heavy Equipment Technicians, a four-year apprentice program with both provincial and interprovincial exams), the consultant assumed they could repair all the equipment to the same level. In other words, the consultant assumed that technicians who repaired backhoes and earthmovers could also repair fire apparatus. A $500,000 hydraulic track backhoe has some complex systems, just like a $500,000 fire truck, but a backhoe is not called twenty-four hours a day, seven days a week to save lives. If the backhoe is not performing to capacity, it merely does not lay as much sewer pipe in a day. The same cannot be said for a fire truck. Had the department invested the money to qualify the fire department technicians to the NFPA 1071 standard, the consultant never would have considered this proposal, and much time and debate could have been saved. As it was, reasoning and cooler thinking prevailed and the merger was rejected.

If your department has full-time technicians, make sure they pass at least the EVTCC exams. In fact, every department could use this as part of a screening test for future applicants. If nothing else, it shows that the applicant has the drive and desire to work on fire apparatus.

Likewise, for the small town department that may have only one staff member who repairs the town's lawn mowers, snow-removal equipment, grader, and fire truck, the technical training described in NFPA 1071 is very important. These people are versatile and flexible by the very nature of their jobs. They usually are also local residents and therefore have a stake in the welfare of their small town. These are the types of people who are eager to improve themselves if given the opportunity. If necessary to justify the training expense, it may be possible to repair fire apparatus equipment from other small towns in the area as a revenue-generating activity.

For the rural department that may have its repairs done at a commercial garage, the technical training is even more important. If the department uses a bid system to decide which garage repairs the apparatus, it should ask to review the qualifications of the people who will be doing the repairs. A shop that has invested the time and money to train its technicians should be proud to include this information in its bid. Remember, however, that having an outside commercial garage actually perform the repairs does not release the department from doing the routine inspections and maintenance required to detect the need for those repairs.

The apparatus maintenance section of the IAFC did not magically appear in 1988, nor was the first fire truck repaired the following day. In fact, many state and provincial associations have been promoting the training of fire service technicians for some time. The California Fire Mechanics Association was formed over thirty years ago and has been training technicians ever since. The Ontario Association of Mechanical Officers has been very successful from its start in 1983. A list of these state and provincial associations with their current addresses and contact information can be found by searching the World Wide Web. The information in this manual represents one more tool that you may use to increase your knowledge and skills as an EVT and, thus, to play your role in promoting the safety of your community. As Frank A. Moffat, technician, educator, and longtime member of the AMS expresses it, "We may not be the pride of the fire service, but without us the pride don't ride!"

The Organization of the Fire Department and the Maintenance Facility

Whether your department serves a small town, a rural farming area, or a large metropolitan area, it must have an organizational chart with clearly defined lines of responsibility and communication. Every person in the department must be familiar with this organizational chart and understand the duties and responsibilities of each individual in the organization. In most cases the technician will not have a direct relationship with the fire department; few departments are big enough to have the luxury of a full-time technician. In many smaller departments the technician will be employed either by the public works department or by a private contractor. Some small departments may be fortunate enough to have a volunteer member as their technician. This type of relationship can be extremely beneficial as the department has a very short line of communication with the technician.

The maintenance facility itself may be physically removed from the fire department. Many large and medium sized cities have repair facilities that are separate from the fire departments that they serve. This is not necessarily a disadvantage, as long as the lines of communications can be kept open. The technicians must be trained to understand not only the unique use of these highly specialized vehicles but also the very different standards of performance for which they are constructed. The fire department must also have very well-trained operators who know not only how to conduct a thorough inspection but also how to interpret the results and communicate them promptly to the technicians.

The Role of the EVT in the Organization

There are two levels of EVTs. The Level I technicians' tasks are to inspect, conduct maintenance, and perform operational checks. These people do not have to be highly qualified mechanics. They could be firefighters with limited mechanical talents but a solid knowledge of vehicle operation. The Level II technician must be able to perform the duties of the Level I technician and also be able to conduct service tests and make repairs. These people must have a solid knowledge of truck construction, operation, lubrication, electrical/electronics, and water hydraulics as they apply to fire pumps. This knowledge must also extend to the differences between fire apparatus and standard highway trucks and to an understanding of the NFPA standards that apply to fire apparatus construction (1901, *Standard for Automotive Fire Apparatus*), maintenance (1915, *Standard for Fire Apparatus Preventive Maintenance Program*), and testing (1911, *Service Tests of Fire Pump Systems on Fire Apparatus*). Armed with this knowledge the EVT can play a vital role in selecting the most appropriate apparatus for any particular department. However, the EVT is often overlooked during the bidding process.

An average department has three main assets: the people, the apparatus, and the buildings. Of course, the value of the people is immeasurable, but the value of the other two can easily be determined, and in most cases the apparatus that sit in the buildings are worth more than the buildings. If a new fire hall were constructed to replace one that was destroyed or out of date, the bidding process would involve at least an architect, a designer, and an engineer. On the other hand, if the apparatus were destroyed, most likely the process of acquiring a replacement would involve the fire chief, the mayor, the town treasurer, and a few other interested and well-meaning people. In most cases none of these people would have any expertise in fire apparatus construction and maintenance. We only talk to the technician when the new apparatus breaks down, and then we wonder why our maintenance program is not working. EVTs therefore need to be involved from the ground floor in choosing a vehicle that meets the unique requirements of the department and complies with all of the specified standards so that they can build an appropriate maintenance program specific to that apparatus.

The Mission of the Fire Service

The mission of the fire service is simple: to save lives and property. Like many things that have a simple definition, however, this mission's application is very complex. Few fire chiefs complain that their firefighters are not brave enough, but many are not happy with the level of maintenance and service that their vehicles receive. In most cases the complaints stem not from the technician's general incompetence but from their not understanding the unique construction and operation of fire apparatus, as opposed to standard over-the-road trucks. If the mission of the fire service is to be fulfilled, the firefighters and EVTs must be not only brave and selfless but also well trained to the appropriate level of the NFPA 1701 standard.

The Fire Department's Standard Operating Procedures, Rules, and Regulations as they Apply to the EVT

While the truck manufacturer will provide a repair and maintenance manual with applicable schedules for oil changes, grease jobs, and the like, each truck will have different requirements, not only because of where it was made but because of where it is used. The maintenance schedule supplied by the manufacturer is only a guideline. A vehicle used in hilly terrain with mostly wild land in a cold climate needs a very different maintenance schedule than a vehicle used in a major city in a warm climate. It does not matter that they may have come from the same factory with the same recommended maintenance manual. If one of its vehicles is ever involved in an accident, the maintenance department must be able to produce at a moment's notice detailed, complete, and accurate records of its repair and maintenance. This is not an option. Therefore, you, as the trained EVT, need to be involved in developing and maintaining a set of standard operating procedures for the inspection, maintenance, and repair of your fire department's apparatus.

Critical Aspects of NFPA 1500 as They Apply to the EVT

NFPA 1500, *Standard on Fire Department Occupational Safety and Health Program,* Section 4-1, Fire Department Vehicles, clearly spells out that the vehicle must be specified, designed, and constructed with the safety of the firefighters in mind. Its acquisition, operation, maintenance, inspection, and repair must always consider the health and safety of the user. For example, far too many firefighters have been hurt or killed by loose equipment in the cabs of vehicles. Any mount used in the cab to hold equipment must be able to withstand a longitudinal force of at least nine times the weight of the equipment being held. This force is referred to as 9-G force. This same mount must also be able to withstand lateral or vertical forces of 3-G. Many well-meaning people do not understand that a flashlight that weighs 5 pounds (2.3 kg) could exert a force of 45 pounds (20.4 kg) during the impact of a crash. A few small sheet metal screws will not be sufficient to withstand such forces.

NFPA 1500, Section 4-2, Drivers/Operators of Fire Department Vehicles, outlines the drivers'/operators' role in safely transporting firefighters to and from the emergency scene. Inasmuch as these drivers/operators are often the people responsible for the inspection of the vehicle, they play a critical role in its safe operation. The International Association of Fire Chiefs reports that one-third of all firefighter deaths occur en route to an emergency. Fire department apparatus drivers therefore must be the best of the best to help reduce this alarming statistic. For their part, the technicians who repair these vehicles must understand the tremendous forces involved in moving apparatus and maintain and repair the fire apparatus so it will function under such mechanical stress.

NFPA 1500, Section 4-3, Persons Riding in Fire Apparatus, outlines the safety standards for persons riding on fire department vehicles. Through many editions, NFPA 1500 has prohibited riding on tailboards, side steps, or running boards. All personnel must be seated and belted while the vehicle is in motion. Many departments have elected to allow their personnel to travel to the fire scene with their SCBA already attached to their backs. In this case the seats must be designed for that purpose. Also a firefighter wearing an SCBA will weigh more than one without. This may seem obvious at first, but if it were obvious, there would not be a standard that says the last step down to the ground from a fire apparatus shall not exceed 24 inches (610 mm). With the weight of the SCBA and the weight of the firefighter, a longer drop could cause back and leg injuries when the firefighter dismounts. As an EVT you must be familiar with these standards to ensure your vehicles' compliance and your firefighters' safety.

NFPA 1500, Section 4-4, Inspection, Maintenance, and Repair of Fire Apparatus, sets out all the standards that must be in place for fire apparatus to be properly maintained and repaired. The firefighter must be trained to conduct a complete and thorough inspection of the vehicle. Any defects must be documented and repaired. In many repair facilities the technicians repair only what is indicated in the firefighter's report. Therefore, if the firefighter does not report a needed repair, it simply will not be made. Ultimately, if the firefighter is not trained to identify problems, the vehicle will not be safe.

NFPA 1500, Section 4-5, Tools and Equipment, specifies that all tools, including ladders, hoses, axes, fire extinguishers, and any other equipment designed for training or fire fighting, shall be inspected at least annually. This task may or may not be done by the firefighters or the EVT. Regardless of who is assigned to the task, it must be clearly outlined and carried out and not left to chance.

Just as any other member of the fire service must wear the appropriate personal protective clothing for the task at hand, so must the EVT. These standards are outlined in Chapter 5 of NFPA 1500. In particular the EVT must wear appropriate eye and face protection. The best means of preventing eye injury is to wear the correct type of eye protection. Many types of safety glasses and shields are available to provide adequate eye protection. Safety glasses provide good protection against metal, dust particles, and other foreign objects. They should be worn whenever the eye is exposed to the risk of being damaged (for example, grinding, cutting metal, using a punch or chisel, or using compressed air to remove dirt). Face shields and goggles also can be used for eye protection. The face shield protects the entire face and should be worn when working with hazardous liquids (chemicals, acids, fuels, or solvents) or flying objects.

Goggles and face shields may be especially useful because they can be worn over regular prescription glasses (most regular prescription glasses are not made of safety glass and lack required side protection). Goggles tend to fit snug against the head for good protection. They are made with clear and colored lenses for nearly every type of eye hazard. Clear lens goggles protect against foreign objects, while colored lens goggles protect against dangerous light rays (oxyacetylene). Special welding lenses are used for arc welding procedures.

CAUTION:
Unless you are using the correct type of eye protection, never use welding equipment or watch someone else welding.

Another hazard of concern to EVTs is the chance of contamination by blood-borne pathogens. Section 4-7 of NFPA 1500 specifies that all emergency vehicles must be decontaminated to satisfy NFPA 1581, *Standard on Fire Department Infection Control Program,* if they have responded to an incident where the potential for contamination by blood-borne pathogens exists. If a vehicle is not properly decontaminated before repairs start, blood-borne pathogens can infect the technicians by entering the body through the eyes or other mucous membranes. The blood-borne pathogens that are of particular concern are HIV and hepatitis A, B, and C. There is a vaccine for hepatitis A and B but not for hepatitis C or HIV. Vaccines are effective only if administered before exposure. Goggles can protect against contamination, but the best protection is to thoroughly clean the vehicle with a one-to-nine mixture of household bleach (one cup of bleach to nine cups of water). Special care needs to be taken in cleaning any sharp corners in cabs or around foot wells that may contain medical waste.

NFPA 1500, Section 5-11, addresses appropriate hearing protection. Equipment such as impact wrenches, air drills, and engines running under a load can cause very loud noises. Exposure to these loud noises for extended periods can lead to hearing loss. Two types of hearing protection may be used. Earplugs, small plastic or rubber devices that are inserted into the ear, should be worn in environments that are constantly noisy. For extremely loud noises, earmuffs can be worn. Earmuffs cover the entire ear and form a sound barrier.

CAUTION:
Your hearing will not recover once it has been damaged.

Many firefighters have suffered hearing loss. You may have noticed that the air horns of modern fire apparatus are located on the front bumpers, not on the cab. This reduces the level of sound in the cab for the firefighter—another example of how fire trucks differ from over-the-road trucks.

Motor Carrier Regulations

Federal Motor Carrier regulations in both the United States of America (Federal Motor Vehicle Safety Standards [FMVSS]) and in Canada (Canada Motor Vehicle Safety Standards [CMVSS]) detail the construction and safety standards of over-the-road highway vehicles, but in most cases the NFPA regulations are tougher. For example, the NFPA standard for air brakes is far more stringent than the FMVSS. If in doubt, err on the side of safety. Although you should never violate an FMVSS, you will find very little conflict between the FMVSS and the NFPA.

Applicable Federal, State (Provincial), and Local Regulations

The regulations that apply to fire department vehicles vary enormously both within and among states (provinces) and from one local jurisdiction to another. In some states the NFPA standards for the building and maintenance of the vehicle have been adopted as law; in other states and provinces they are used only as a guideline. In one Canadian province the fire trucks do not even have license plates, nor do the drivers need a license of any kind to operate them. More progressive states and provinces require that the operator have not only a driver's license for comparable commercial trucks but also a working knowledge of the brake system. Operators in other states are exempt from commercial driving license (CDL) requirements. However, the fire chief standing in front of a jury must find it very difficult to explain why the person driving the fire truck at a great rate of speed did not need the same level of knowledge of the air brake system as the person driving the city dump truck.

Manufacturer's Specifications, Inspection Checklists, Maintenance Schedules, Maintenance Checklists, and Department SOPs

How a particular department interprets and uses manufacturer's specifications, inspection checklists and maintenance schedules, maintenance checklists, and department SOPs will depend upon that department's overall maintenance strategy and philosophy. A department's maintenance styles will fall into one of three major levels: crisis maintenance, preventive maintenance, or predictive maintenance. Only when a department recognizes where its present maintenance style falls can it move on to the next level.

The first level, and the one far too many departments use, is *crisis maintenance.* It is also called run to failure and involves a management style that does not believe in long-range planning. Around this type of department you will hear phrases like, "If it's not broke, don't fix it," "Don't go looking for trouble," or "It's never done that before" (the last, simply because no records have been kept and no one can remember the last time the apparatus broke down). This level of maintenance has low administrative costs, if you don't count the time spent blaming other people for your problems. While it may be resourceful to be able to fix anything with a piece of hay wire, a spring from a ballpoint pen, and a roll of duct tape, you are in big trouble if this is your department's maintenance style.

In most cases, using a crisis style of maintenance is not a planned decision. It quite often just evolves over time. During the 1980s, fiscally responsible jurisdictions made a big push to do more with less. This was in response to a general societal perception of far too much waste in the public sector. In reality, instead of doing more with less, many departments simply did less with less. This happened at the same time that electronics and the first generation of computers for engine and transmission control became available. Many administrators thought that computers would make the technician's task easier. So at a time when they actually needed more technicians trained to a higher level, many departments cut staff. Also many cities consolidated the fire service mechanics with the city fleet mechanics in an attempt to save money. This cost-cutting destroyed the bond of trust that had been created between the fire service mechanics and the firefighters whose lives depended on them. This close bond had helped to ensure that the lines of communication were kept short and apparatus were kept in a high state of readiness.

If you spend most of your maintenance day running from one crisis to another, your new equipment will soon become unreliable and the firefighters will lose faith in the equipment's abilities to function. If this happens your department will very quickly begin to fragment, with maintenance in one corner, firefighters in another, management in the third, and soon the public in the fourth. When the public gets involved, your department will be under a microscope, and people will start to spend their energies blaming one another instead of searching for solutions. As long as you are in the crisis mode, you will never have time to move to the next level of maintenance. Having said this, there are times when a department does have to make a conscious decision to use the run-to-failure mode. It would be far better to run an engine to failure because of an overheating problem than to shut it down sooner, cut off the water supply, and fail to get a firefighter out of the building.

The next level of maintenance is called *preventive maintenance*. This is time-based in that a schedule fixed on number of hours, gallons/liters of fuel, or miles/kilometers is followed for apparatus maintenance and testing. The routine equipment inspections at this level of maintenance can prevent some failures by requiring technicians to disassemble or open the machine to inspect it. Preventive maintenance is based nonetheless on a manufacturer's manual that may or not apply for the fire service. For example, at the conclusion of a seminar I conducted on 0W-40 motor oil and why it was a better motor oil for the fire service than the presently used 15W-40, a student asked me whether I realized that all the major diesel engine manufacturers' manuals recommend 15W-40. My reply was that these are the same manuals that recommend that the engine be up to operating temperature before being put under full load, a luxury rarely afforded fire apparatus. Some parts of the manufacturer's manual apply to a fire service vehicle, and some do not. Less than one percent of all trucks made in North America are made into fire apparatus, and not enough is being done to understand the complete life cycle of these vehicles.

Even if you choose to follow a preventive maintenance program as outlined by the manufacturer, you will still have breakdowns. These breakdowns will be unplanned and expensive; apparatus will fail at inconvenient times. Management will quickly turn against the repair department if you have led them to believe you could cut costs if only you had a preventive maintenance program: wrong. This is one of the great myths about a preventive maintenance program.

Here are a few reasons why a preventive maintenance program might not live up to your expectations. First, your department probably used crisis maintenance as its start point. It is impossible simply to switch from crisis maintenance to preventive maintenance and assume that all will be well. In fact, some of your equipment may be too badly worn to be repaired and will just have to be scrapped. Second, many preventive maintenance programs are not based on a fire-service application but on an over-the-road truck application. High heat, large electrical demand, wide-open throttle (WOT), cold startups, and long periods of engine idle make your maintenance standards very different from those of other applications. Third, a preventive maintenance program prepared for a particular truck by one department cannot necessarily be used for the same truck in another department. All programs must be custom designed for each piece of equipment, taking into consideration its operational use and the level of maintenance skills in the department. Such concerns as terrain, temperature, humidity, road speed, and the operators' level of training must be considered. Unless all of these are taken into account, your preventive maintenance program will be a failure, and the money and time spent on it will be deemed a waste. Further, your department could easily fall back to a crisis maintenance program. A good preventive maintenance program is designed to prevent failures through regular routine inspections that are recorded accurately and in detail. Record keeping, however, is expensive and often the first thing to go during a downsizing. While computers can help to manage this information, they are still only as good as the information put into them.

The highest level of maintenance is *predictive maintenance;* it is also sometimes called proactive maintenance. Simply put, for this level of maintenance you do every thing that you would do for a good preventive maintenance program, but in addition you ask this simple question: "Why?" Every time you perform a maintenance routine, you ask, "Why?" For example, I often ask departments, "Why do we change oil every 250-engine hours?" I get all kinds of different answers ranging from "That's the way we have done it since the 1950s" to "The manual for my tractor engine says so and it's the same engine as in the fire truck" or "That's when we grease the chassis, so let's do it all at once."

Why do we change oil every so many hours? Why do we use that type of grease? Is oil filter X better than oil filter Y; why or why not? The simple questions are often the hardest to answer. Even with the best of maintenance, parts will fail. Predictive maintenance is based on the premise that you must be able to predict *when* they will fail. For example, a starter for a diesel engine is good for about 20,000 starts. If these starters were used in a delivery truck and you knew by looking at the route that the unit would make 100 pickups in the mornings and the same in the afternoon, the starter would be good for 16 six-day weeks. Under a predictive maintenance program, at the end of the 16 weeks or sooner, the unit would come into the shop and the starter would be changed out. Yes, the starter would be changed when it was still working. Predictive maintenance changes things before they break down.

To be able to predict a failure you must have a very clear understanding of what causes the failure in the first place. This is called failure analysis.

Parts fail for many different reasons. My experience in the fire service has revealed the following failure causes:

- Five percent are caused by operator error. People do back over things and run into curbs. Some mechanics even say that nothing can ever be made completely firefighter proof.

- Fifteen percent are caused by the nature of the job — high under-hood temperature, full-throttle accelerations with a cold engine, larger than normal electrical demands.

- Fifteen percent are caused by obsolete equipment — it has simply run its life cycle and that's all there is to it. The fire service is unreasonable to expect a high performance truck to last 20 to 25 years. Metal fatigue, rust, and corrosion will eventually take their toll.

- The remaining 65 percent are caused by lack of maintenance. This 65 percent can be broken down farther: over 50 percent of these failures are due to lack of proper lubrication. In other words, if you want to move to a predictive maintenance style you must understand lubrication, or what is now more commonly called tribology. (*Tribology* comes from the Greek word *tribos,* meaning "to rub." It was coined in 1966 by H. P. Jost to describe the interrelated sciences of friction, lubrication, and wear.) A good predictive maintenance program uses oil analysis, heat analysis, and vibration analysis, along with an excellent tracking system (sometimes called bookkeeping).

Will predictive maintenance save money? Not really. At the beginning, it will be more expensive in hard costs (tires, batteries, and starters), but in the end it will save in the costs that are hard to measure, such as lifesaving reliability and public goodwill.

If you are going to build a maintenance program, you have to do it almost from the ground up. The suggested program in the vehicle's manual is just that, suggested; you will have to tailor your own program from it. How do you begin to build your own customized predictive maintenance program? By using some predictive maintenance technologies such as oil analysis, vibration analysis, heat analysis, and most important of all, root cause analysis. While using other departments' schedules and other people's information gained from trial and error may be convenient, your department's needs will be different from those of any other department.

Tool and Equipment Safety

Mechanic work in general requires many expensive and specialty tools. Fire apparatus repair needs even more. If your shop chooses not to purchase and train its personnel in the uses of these tools, then take your apparatus to a shop that has them.

Accidents are often caused by carelessness resulting from a lack of experience or knowledge. Make sure that you use the proper tool for the job and use it the right way. The wrong tool or the right tool used incorrectly can damage the part on which you are working or cause injury — or both.

Hand Tool Safety

Improper use of hand tools has caused many accidents (use the correct tool for the job). A good technician keeps shop tools and workbenches clean and tidy. Oil or grease on tools can cause serious personal injury. Keep all hand tools grease-free and in good condition. Greasy tools can slip, fall into moving parts, and fly out causing serious injury. Keep your hands as clean as possible when handling tools.

Take care when using sharp or pointed tools. Do not carry screwdrivers, punches, or other sharp objects in your pocket. You could injure yourself or damage the unit on which you are working. Make sure the tools you are going to use are sharp and in good condition. Do not use broken or damaged tools.

Power Tool Safety

Compressed air, electricity, or hydraulic pressure is used to operate power tools. Always wear safety glasses or goggles when using power tools. Do not operate power tools or equipment that is in an unsafe condition. Make sure electrical cords and connectors are in good condition. Never use electric power tools on a wet or damp floor. Never leave a power tool unattended while switched on. When you have completed the job, turn the power off. Do not talk to or distract anyone who is using a power tool.

Bench grinding wheels and wire brushes should be replaced if defective. The grinder has a guard around the wheel to prevent injury if the wheel should explode. Never use the grinder if this guard is removed. The bench grinder also has a safety shield that protects the operator from flying chips or abrasives. The tool rest on a bench grinder prevents the object being ground from being pulled between the wheel and the guard (Figure 1-1). As the grinding wheel wears, the space between the wheel and the tool rest increases. The tool rest must be adjusted so that the clearance between it and the tool rest is no more than $\frac{1}{4}$ inch (6 mm) (Figure 1-2). Be sure that the grinding wheel is rated (rpm size) for the specific machine. Too many rpms could cause the wheel to explode. A photo tachometer is a handy tool to check the rpms of power tools (Figure 1-3). Grinding and cleaning require skill and careful handling to avoid injury to the operator or the tools and parts being reworked.

CAUTION:
Always wear eye protection when using a grinder.

Figure 1-1 Grinder. Note the safety face shield and ventilation pipe behind the face shield. Not shown in this picture is an incandescent light above the grinder. Never use a fluorescent light above any rotating piece of equipment.

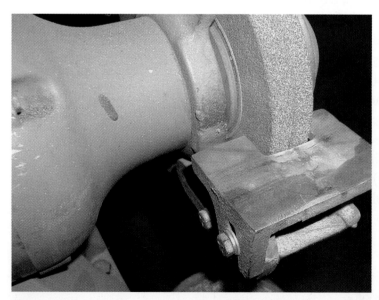

Figure 1-2 Grinder. Note the space between the grinding wheel (stone) and the work rest; this distance should never exceed ¼ inch or the piece being ground can get trapped and fly up.

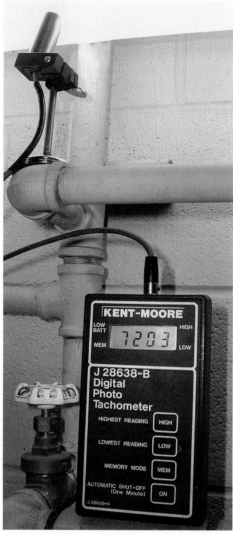

Figure 1-3 Photo tachometer. In this picture, the bright light at the top is a fluorescent light and just below that is the sensor for the photo tachometer. Because the light is powered by a 60-cycle AC power source, the photo tachometer will read very close to 7,200 rpm. It is for this reason that a fluorescent light should never be placed above a piece of rotational machinery, such as a drill, saw, or grinder. If the machine rotates at a speed of 7,200 rpm or a multiple or factor of 7,200 rpm (e.g., 3,600, 1,800), the machine will look like it is not turning at all. This would be very dangerous. The effect of the fluorescent light on your perception of the rotation of the machine is similar to what occurs when a technician uses a strobe light to adjust the timing on a gasoline engine: the vibration dampener/front pulley appears to be stopped. The photo tachometer also is useful for checking the speed of engines and electrical motors such as pump primers or pump shafts.

The hydraulic press should not be used beyond its rated capacity. Most hydraulic presses have a pressure gauge to indicate how much pressure they are exerting on a part. All work should be properly positioned and supported, and all recommended shields and protective equipment should be used. The press develops extreme pressure (several tons) that can cause parts to explode, causing serious injury.

Compressed Air Safety

CAUTION:

Pneumatic tools must always be operated at the pressure recommended by the manufacturer.

Tools that use compressed air to operate are known as pneumatic tools. Compressed air is used to inflate tires, operate air tools, spray paint, and blow parts dry. Be careful when using compressed air to blow away dirt from parts. Always use eye protection when using compressed air.

Before using compressed air, check all hoses and connections. A badly worn air hose could burst under pressure. Do not use compressed air to clean your clothes. This can cause dirt particles to be embedded in your skin and cause infection. Do not point the air nozzle at anyone. If the air nozzle is directed at the skin, an air bubble can be injected into the blood stream causing serious injury. Avoid the use of compressed air to blow off parts whenever possible. Place parts washed in solvent on a clean rag and allow them to air dry. Never spin bearings dry using compressed air. If the bearing is damaged, the balls or rollers may fly out causing injury.

CAUTION:

If a bearing is spun dry by compressed air, there will be no lubrication between the rollers and cup, and the bearing may seize due to heat and friction, thus causing serious injury.

Lift Safety

Always take care when raising a vehicle. Never jack a vehicle when anyone is under it. Position the jack so that its wheels can roll as the vehicle is lifted. Otherwise, the lifting plate may slip on the frame, or the jack may tip over. Always use wheel blocks when using a jack to raise a vehicle and place them both in front of and behind the wheels. They will prevent the vehicle from rolling forward or backward.

A vehicle should always be supported on jack stands when it is jacked up for service. Use jack stands only on a level concrete surface and be sure to place the jack stands in the recommended locations on the frame. Always use the correct rating of jack stand. The jack should be removed after the jack stands are positioned safely and securely. Do not leave jack handles lying across the floor in the down position. Someone could trip on them and be seriously injured or cause the vehicle to fall. Never use a jack by itself to support a vehicle; jacks have failed and vehicles have fallen and crushed people.

CAUTION:

Never use a jack to lift an object heavier than it is designed for.

Crane and Chain Hoist Safety

The hydraulic floor crane (cherry picker) and the chain hoist (chainfall) are both used for lifting very heavy loads. The crane has wheels so that the heavy component can be removed or positioned in the proper location (Figure 1-4). The chain hoist may or may not have wheels and is not as convenient as the crane because it is harder to maneuver an engine block into position during installation. When using a chain hoist, be careful not to let the chain damage the paint on the vehicle.

An engine sling or lift cable is used to hook the crane to the engine block. The sling must be tightened against the engine block or inserted into the lifting eyes bolted in place on the engine to facilitate installation

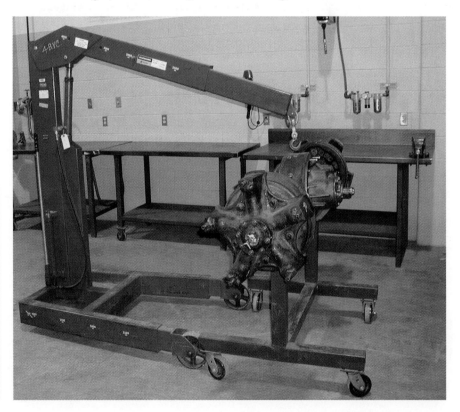

Figure 1-4 Floor crane. Ensure that the load to be lifted does not exceed the lift rating of the crane and that it does not cause the crane to tip over.

Figure 1-5 Transmission stand being used during a transmission rebuild. The stand ensures that the transmission is balanced and easier to work on. This same stand with different adapters can be used for small-block gasoline engines. Large-block gasoline engines and diesel engines will need a larger stand.

and removal. For proper balance place the crane and sling directly over the assembly to be raised. The crane and sling chosen must be rated for the intended load.

Once an engine has been removed safely from its equipment, it should be mounted into an engine stand (Figure 1-5). Most engine stands allow for rapid disassembly of any engine because the engine can be easily rotated through 360°. A good engine stand will have quality casters and brakes to allow the technician to safely and easily move it around the shop floor yet quickly secure it in a stationary position.

CAUTION:
Never work on an engine that is hanging from a hydraulic lift (cherry picker).

Fastener Safety

A number of cases of counterfeit bolts have been reported in recent years. These bolts do not meet their stated design strength and will fail. Never reuse a lock washer; it's not worth it. NFPA 1901, 11-2.1, also clearly forbids the use of star washers on electrical circuits.

Maintenance Equipment Safety

The maintenance equipment needed for the various systems of the fire apparatus will be discussed in the chapters devoted to each particular system. Two excellent texts to supplement the material in this text are Andrew Norman and Robert Scharff, *Heavy Duty Truck Systems* (Albany, N.Y.: Delmar, 1997) and Robert N. Brady, *Heavy-Duty Trucks: Power Trains, Systems, and Service* (Columbus, Oh.: Prentice Hall, 1997).

Workplace Safety

Just as safety is the number-one consideration when firefighters respond to an emergency, so must it be when EVTs work on the fire apparatus. Nothing looks worse than when a fire hall burns down, which could be one consequence of ignoring workplace safety practices. Similarly the safety of the firefighters traveling to and from the fire depends upon the reliability of the fire apparatus and hence the safety practices of the EVT.

Work Area Safety

Familiarize yourself with the way that the shop is laid out. Find out where things are in the shop. You will need to know where the shop manuals and department SOPs are kept in order to obtain specifications and service procedures. Find out whether certain stalls are reserved for special jobs. Take note of all warning signs around the shop —no-smoking signs, special instructions for shop tools and equipment, and danger zones. These signs are there to help make the shop run efficiently and safely.

Never run an engine without proper ventilation and adequate means of getting rid of exhaust gases. Exhaust gas contains deadly carbon monoxide (CO), which can and will kill. Never stand directly in front of or behind a vehicle when it is running and someone is in it. Be sure to set the emergency brake in the on position when running a vehicle regardless of whether the transmission is in park (for an automatic transmission) or neutral (for a manual transmission).

Clothing Standards

Wear proper clothing. Loose clothing, ties, long hair, and jewelry can get caught in rotating parts or equipment and can cause injury. Never wear rings, watches, bracelets, or neck chains. Always wear safety shoes in the shop. Wear shoes with steel-tipped toes and nonskid soles. These shoes provide protection over the toes to prevent injury if a heavy object falls on your foot. Always wear approved rubber gloves when handling dangerous chemicals and acids.

Breathing Protection

Many hazardous materials and chemicals are used in a repair shop. Brake and clutch work can be hazardous because asbestos is used in the manufacture of these components. Breathing asbestos dust can cause damage to your lungs. Breathing protection is provided with respirators. Two types of respirators can be used. The more common is a filter type that works to clean the air going into the lungs. The other type of respirator supplies clean air for breathing. Never blow the dust off brake or clutch parts because this can cause the asbestos to fly into the air. Use approved types of brake cleaning solvent or a special brake vacuum cleaner.

CAUTION:

Asbestos is a harmful material found in brake and clutch dust. It has been found to cause lung cancer.

Lifting and Carrying

The back is one of the most frequently injured parts of the body, usually because of improper lifting. Always lift and work within your ability. Lifting even small, light objects causes some injuries. If people twist their bodies while lifting objects or lift when the load is unbalanced, serious injury can result. To prevent injury to your back, use correct lifting techniques allowing your legs to take the strain, not your back.

In the event of an accident, be sure to inform your instructor or supervisor. These people know what procedure to follow to ensure your safety.

Fire Safety

Make sure that you know the exit route in case of fire. Note the location of fire extinguishers around the shop (Figure 1-6). Take time to read their operating instructions and the type of fire on which they are meant to be used. Fires are classified and controlled according to the types of materials involved.

- *Class A:* Class A fires involve ordinary combustibles such as wood, paper, and cloth.

- *Class B:* Class B fires involve flammable and combustible liquids and gases such as gasoline and propane.

- *Class C:* Class C fires involve active electrical equipment. Note that once the electrical power is shut off, the fire becomes an A, B, or D fire, depending on what is burning.

Figure 1-6 Fire door. This picture shows a fire extinguisher sign well above the shop floor. This is necessary so that you will know where the extinguisher is located if a large truck is in the shop bay. The fire alarm is beside the door with an emergency shutoff switch just below it. Instructions in case of fire are clearly posted on the yellow sheet above the fire extinguisher.

- *Class D:* Class D fires involve combustible metals such as magnesium.

- *Class E:* Class E fires involve nuclear material, nuclear fusion.

- *Class K:* Class K fires involve commercial kitchen cooking media.

Multipurpose dry-chemical fire extinguishers are the most common type of extinguishers used in repair shops. They will extinguish ordinary combustibles, flammable liquids, and electrical fires. Fire extinguishers must be checked periodically (according to local regulations) to make sure they are properly filled and in good working order.

CAUTION:
Never use water on a flammable liquid fire. Water will just spread the fire.

Other Personal Safety Rules

Safety is the responsibility of everyone in the shop. Never take part in horseplay or practical jokes. They can lead to injury. Take note of all the warning signs around the shop. Never smoke while working on a vehicle or while working with any machine in the shop.

First Aid

Make sure that you are aware of the location and contents of the first-aid kit in your shop. If your shop has any specific first-aid procedures, make sure that you are aware of them and follow them. You should be able to quickly locate emergency telephone numbers such as doctor, ambulance, police, and fire department. Your shop should have an eyewash station where you can rinse your eyes thoroughly should you get acid or some other irritant in them (Figure 1-7).

Burns should be cooled by immediately rinsing with water and then treated according to standard first-aid procedures. If someone becomes poisoned by carbon monoxide (CO), transfer the person to a fresh air area immediately. In the case of a severe bleeding accident, try to stop the loss of blood by applying pressure with clean gauze on or around the wound. Call for medical aid as quickly as possible. Do not move individuals who may have broken bones except to remove them from life-threatening danger. Moving a person in this condition could cause additional injury. Call for medical assistance.

Cleaning Products and Procedures

Cleaning parts is an important aspect of every repair procedure. Several types of chemicals and cleaning equipment are found in most repair shops.

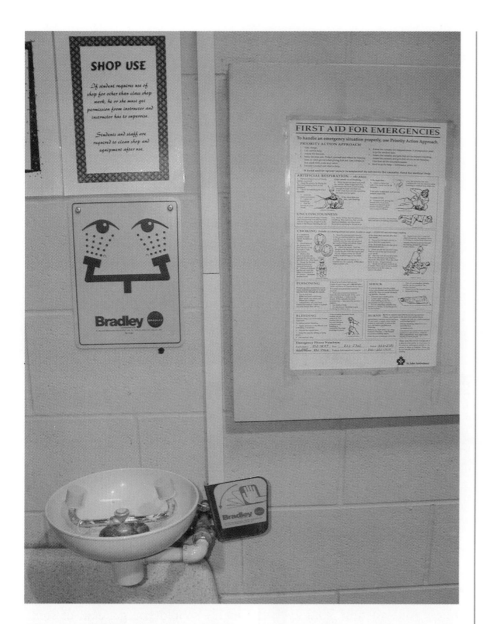

Figure 1-7 Eye wash station. Note also the posted first-aid instructions.

All states and provinces currently have in place regulations governing the disposal of hazardous chemicals. You should be familiar with all local laws and regulations controlling the use and disposal of any cleaning chemicals used in the shop.

CAUTION:
Substantial fines are levied against companies and individuals who fail to follow responsible disposal practices.

Some alkalis, detergents, and solvents can irritate the skin or be harmful to the eyes. When using cleaning chemicals, wear eye protection, nitrite gloves, and coveralls to prevent exposure to the skin and eyes. Ensure adequate ventilation when working with cleaning chemicals.

Use extreme care when spraying cleaning chemicals to prevent injury to other personnel and to avoid an accident. Steam cleaning uses hot water vapors mixed with a chemical cleaning agent to remove dirt and grease from an object. The runoff from this process must be collected and contained. It cannot be allowed to flow into a public sewage system, as it is an environmental hazard. After steam cleaning any object, thoroughly flush all parts with clean water and blow them dry with compressed air.

Become familiar with any cleaning equipment in your shop, such as high-temperature washers or other parts washers (Figures 1-8 and 1-9). High-temperature washers use a soap solution at high temperature to wash large engine and transmission parts. They replaced washers that used caustic soda, which is very expensive to dispose of properly at waste treatment centers. The sludge in the bottom of high-temperature washers will also have to be disposed of correctly; do not pour it down the sewer.

Figure 1-8 High temperature washer.

Figure 1-9 Parts washer. These parts washers do not use a VARSOL solution. They use a warmed soap solution that is nonflammable. Protective gloves should be used.

WARNING:
Never place aluminum parts in a caustic soda washer: the aluminum will dissolve.

Cold-tank solvent cleaning can be used for cleaning steel and aluminum parts. The solvent softens and washes away grease, oil, and sludge. When using a solvent as a cleaning agent, you must be cautious that it is not absorbed through your skin and into your body. Wear nitrite gloves and suitable coveralls when working with solvents. Solvents are flammable liquids; therefore, you should not have an open flame near a solvent tank.

Bead blasting (sand blasting) is another cleaning method used in apparatus maintenance (Figure 1-10). This method relies on physical abrasion to clean the surface. A bead blaster produces a very fine dust; therefore, you should wear a dust mask when placing parts in or removing them from the cabinet. All traces of sand must be removed from any component that has been cleaned in a sand blaster before it is used because even very small particles of sand will ruin an engine or transmission rebuild.

CAUTION:
Never use gasoline or paint thinner as a cleaning agent.

 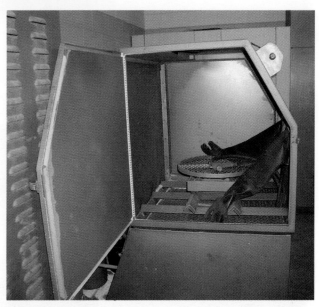

Figure 1-10 a. Sandblaster with the cabinet door closed. **b.** Sandblaster with the cabinet door open. Note the rubber gloves that protect the operator. All traces of sand must be removed from any component that has been cleaned in a sand blaster before the component is used because even very small particles of sand will ruin an engine or transmission rebuild.

Another type of cleaning method relies on heat to break off or oxidize the dirt. This type of cleaning leaves an ash residue on the surface that must be removed by spray washing (Figure 1-11).

Housekeeping
Workplace accidents rarely result from a single cause. More often than not, many factors contribute to an accident. However, very often one of those factors is a disorganized shop. A clean, well-organized shop is important to the safety of technicians. Keep the shop floor and workbenches clean and tidy. Oil, grease, or coolant on the floor can cause serious personal injury. Oil and coolant spills should be cleaned up immediately with a

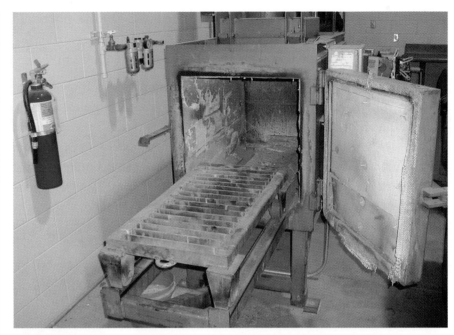

Figure 1-11 High temperature engine cleaner. This device uses very high heat to burn off the grease on an engine block. The residue left is a very fine ash that will need to be removed. It can take hours to heat and cool a large engine block. Make sure that any bolts or studs are removed from the engine block before heating. Note the extinguisher on the wall.

Figure 1-12 Airtight garbage can for the disposal of oily rags. Oily rags must be disposed of in an airtight can as they can spontaneously combust if the temperature and humidity are right. The can must be emptied every night or every twenty-four hours.

Figure 1-13 Fuel station used to store small amounts of fuels inside the shop. Ensure that no welding or grinding can occur close to this cabinet. Note the lock on the cabinet. The plastic fuel container is normally stored inside this cabinet, but it is not safe to do so because the top is missing from the spout. The top must be found and replaced before the container can be stored in the cabinet. A fire extinguisher properly sized for the amount of fuel being stored must be located within a safe distance to serve this fuel cabinet.

commercial absorbent. Grease should be wiped up and then cleaned with commercial absorbent. Dirty and oily rags should be stored in closed metal containers to avoid catching fire (Figure 1-12). Never allow an open flame near flammable solvents, chemicals, or battery acid. Combustible materials must be stored in approved and designated storage cabinets or fireproof rooms (Figure 1-13).

Hazardous Materials

Most shop chemicals can be dangerous to your health. Some of them may irritate the skin. A material is considered hazardous if it causes illness, injury, or death or pollutes the water, air, or land. Cleaning chemicals, used oil, and some types of engine coolants are some of the hazardous substances with which technicians come in contact. Solvents and cleaning chemicals used in the shop must carry warning and caution information that must be read and understood by everyone who uses them.

In Canada and the United States, environmental protection agencies (EPA) regulate the disposal of hazardous waste. A material is not considered to be hazardous waste until the shop is finished using it and ready to dispose of it. The shop is responsible for the safe disposal of hazardous wastes (Figure 1-14). Licensed waste disposal companies will dispose of the shop's waste carefully and safely. It is imperative that you follow all federal, state or provincial, and local laws dealing with hazardous materi-

Figure 1-14 Used oil container. This inside container replaces an outdoors-underground used oil storage tank. Used oil is a hazardous waste because it is carcinogenic; always wear rubber gloves when handling it. This container also provides the means for draining buckets and used oil filters before they are disposed of.

als (dangerous goods). You are responsible for keeping current with these rules as they can and do change rapidly. Your department may be exempt from some of these regulations, but on the other hand, you may be held responsible for any waste you generate even after it has been hauled away. Make sure you use a reputable waste disposal company and follow practices that generate less waste.

All workers must also be familiar with the Workplace Hazardous Materials Information System (WHMIS) and must have access to material safety data sheets (MSDS), which provide dangerous goods information. The purpose of WHMIS is to inform you and fellow workers about hazardous materials you may find on the work site. It is mandated by federal and provincial legislation that became effective on October 31, 1988, recognizing workers' right to know about hazardous materials in the workplace. WHMIS will provide information and warnings about hazardous materials with which you may come in contact. Emergency response groups also use this system to develop safe practices and emergency measures.

Hazardous products are classified into three categories: restricted products, prohibited products, and controlled products. WHMIS applies only to controlled products, not to restricted or prohibited products. Furthermore, it applies only in the workplace, which is defined as any location that is in or near where people are working. It does not apply in domestic situations, such as in homes.

WHMIS consists of four major elements: labeling, material safety data sheets, worker education, and protection of trade secrets. All hazardous materials in the workplace must be clearly labeled (Figure 1-15). WHMIS legislation states that labels must have a specific content and design, and it specifies two basic types of labels:

1. *Supplier labels:* these must be attached to containers of hazardous products when they enter the work site.

2. *Workplace labels:* these are placed on containers of hazardous products that do not have supplier labels (for example, WD-40).

Material safety data sheets (MSDS) must be provided for all hazardous products in all workplaces (Figure 1-16). An MSDS contains very detailed information about a single product.

WHMIS protects the suppliers as well as the users of hazardous products. It allows suppliers to withhold certain information about a hazardous product in order to protect trade secrets from competitors; however, suppliers cannot withhold information that is required for the users' safety. If a supplier chooses to withhold information about a product, it must have approval from a screening agency.

Suppliers are responsible for classifying their products, developing labels, and developing MSDS. The employer is responsible for ensuring

Figure 1-15 MSDS label. Anytime a hazardous material is transferred to another container, a workplace label must be made for the new container. The container depicted here never had a supplier label; therefore, even though the material is in the original container, it needs a label. Any new container should come with a supplier's label. This information is posted in the MSDS book for all to see.

Figure 1-16 MSDS information center. Material safety data sheets must be available to all workers. The sheets must be updated regularly. Emergency procedures must be clearly displayed. MSDS sheets should be posted in a high traffic area but should not be posted next to a worker's bulletin board where they might be covered.

that information from suppliers is available to employees. Employees in turn are responsible for knowing how WHMIS works.

The Occupational Safety and Health Administration (OSHA) is a U.S. federal agency in charge of workers' safety. This organization develops standards and conducts inspections to ensure the workplace is a safe and healthy environment. OSHA's Hazard Communication Standard, developed in 1983, requires chemical manufacturers to publish information about the chemicals they produce and distribute. Canada adopted many OSHA standards when developing its workplace safety regulations.

Every employee in a shop is protected by the right-to-know law concerning all chemicals. All of the information about each chemical must be posted on a material safety data sheet (MSDS) and must be accessible to each employee. The manufacturer of a chemical provides these sheets upon the customer's request at the time of purchase. The MSDS includes detailed information about the product's chemical composition and precautionary information that can prevent a health or safety hazard.

The employer must provide hazardous materials training to each new employee and to all employees annually. This training will help the employee become familiar with the use of required protective clothing, cleanup procedures (in case of a spill), and any other information regarding the safe handling and use of any hazardous materials.

The employer must provide labels on hazardous materials including information about the product manufacturer's name and address, the product's chemical name, and safety information about the product. A typical label may document whether the material is poisonous or flammable, whether wearing protective equipment is required, and whether the chemical poses a reactivity hazard. Employers also must maintain proper documentation on the hazardous chemicals used in the shop, proof of employee training, and accessibility to all MSDS for any hazardous chemicals used in the shop.

The manufacturer of hazardous waste materials must provide all warnings and precautionary information that must be read and understood by the user. Many people have been injured when they were unaware of what was in a container. This is common when people transfer chemicals or materials from one container to another. Always make a new label for the new container because you may not be the only person to use this chemical or material.

CAUTION:
Unlabeled materials can be very dangerous.

Summary

Whether or not you are a member of the fire department, as the EVT responsible for the inspection, maintenance, and repair of fire department apparatus, you need to be aware of the fire department's mission, its organization, and its function so that you will understand your role in the system. This chapter has described the general knowledge requirements for an emergency vehicle technician as outlined in NFPA Standard 1071. This should supply the basis for you to function within and contribute to the safe operation of the fire department.

Chapter 1 Review Questions

Short Answer

Write your answers to the following questions on the blank lines provided.

1. Do NFPA standards define minimum or maximum requirements?

   _____

2. What are the four most important questions a fire chief must ask when deciding on a maintenance program for the department's equipment?

   _____
   _____
   _____
   _____

3. NFPA 1071 was designed for big city, full-time departments only. True or false?

   _____

4. Briefly explain the difference between the two levels of EVT as defined by NFPA 1071.

   _____
   _____
   _____
   _____

5. Should the EVT be involved in the fire apparatus bidding process? Explain your answer.

   _____
   _____

6. Why don't maintenance manuals made for engines and transmissions designed for over-the-highway use apply to fire apparatus?

   _____
   _____

7. Why should each department develop its own maintenance schedule?

   _____
   _____
   _____

8. Name the three common styles of maintenance and specify which one your department embraces.

   _____
   _____
   _____

10. Which style of maintenance is time based, and why might it not work for your fire department?

   _____
   _____

11. What is the other often used name for the predictive maintenance style?

_____

12. When are parts changed in a predictive maintenance program?

_____

13. What percentage of your own department's time is spent repairing failures caused by each of the following: operator errors, nature of the job, obsolete equipment, and lack of maintenance? Why is it important that you carry out this type of assessment?

_____

_____

_____

_____

14. How much money does your department spend to repair the failures due to each type of cause listed in question 13. Why is it important to make this calculation?

_____

15. Why is detailed record keeping so important to the effective maintenance of the fire trucks?

_____

_____

_____

16. In addition to NFPA 1071 and 1915, what other standards should be consulted to produce a maintenance program?

_____

17. What is a CDL, and do you have one?

_____

18. Why should the grinding wheel be close to the work rest?

_____

_____

19. Why should fluorescent lights not be used over a piece of rotating machinery?

_____

_____

20. The strength of a bolt is very important. Place in order from lowest strength to highest: grade 8, grade 12, and grade 5.

_____

21. Why must oily rags be disposed of in an airtight container?

_____

22. What do the letters WHMIS stand for?

_____

23. What information does a material safety data sheet contain?

_____

_____

24. Can a manufacturer withhold information regarding worker safety from a WHMIS sheet?

_____

25. Who is responsible for knowing about WHMIS and material data sheets?

_____

Multiple Choice

Write the letter of the correct answer on the blank provided.

_____ 1. Technician A says compressed air is dangerous. Technician B says that compressed air can be used to blow off debris from clothing, skin, or hair if used cautiously. Who is correct?

    a. Technician A

    b. Technician B

    c. Both A and B

    d. Neither A nor B

_____ 2. Technician A says that used antifreeze can be properly discarded by pouring it down the shop drain. Technician B says that gasoline must be properly stored in a safety container. Who is correct?

    a. Technician A

    b. Technician B

    c. Both A and B

    d. Neither A nor B

_____ 3. Technician A says that an MSDS manual contains information about hazardous chemicals. Technician B says the MSDS refers to fire and explosion hazards only. Who is correct?

    a. Technician A

    b. Technician B

    c. Both A and B

    d. Neither A nor B

_____ 4. Technician A says that carbon monoxide can be detected by its distinct odor. Technician B says that carbon monoxide can be easily detected by its gray color in the air. Who is correct?

    a. Technician A

    b. Technician B

    c. Both A and B

    d. Neither A nor B

_____ 5. Technician A says the purpose of WHMIS is to inform workers about hazardous materials they may find on the work site. Technician B says that WHMIS applies only in the workplace and not in the home. Who is correct?

a. Technician A

b. Technician B

c. Both A and B

d. Neither A nor B

_____ 6. Technician A says that a dry-chemical type fire extinguisher is multipurpose and can be used to extinguish all classes of fires. Technician B says fire extinguishers need to be recharged and checked periodically. Who is correct?

a. Technician A

b. Technician B

c. Both A and B

d. Neither A nor B

_____ 7. Technician A says that a good quality screwdriver may be used as a pry bar. Technician B says that a good quality electric power tool cord requires no ground. Who is correct?

a. Technician A

b. Technician B

c. Both A and B

d. Neither A nor B

_____ 8. Technician A says the right-to-know legislation deals with the hiring and firing procedures of a shop. Technician B says the right-to-know law deals with emissions standards. Who is correct?

a. Technician A

b. Technician B

c. Both A and B

d. Neither A nor B

_____ 9. Technician A says that an employer is responsible for training employees about hazardous materials used in the workplace. Technician B says that in accordance with OSHA, employees are responsible for training themselves. Who is correct?

a. Technician A

b. Technician B

c. Both A and B

d. Neither A nor B

_____ 10. Technician A says that jack stands can be used to support a vehicle even if the rated capacity of the stands is exceeded. Technician B says that you may use a jack stand safely if its rated capacity is not exceeded by more than 10 percent. Who is correct?

a. Technician A

b. Technician B

c. Both A and B

d. Neither A nor B

Chapter 2
Types and Construction of Fire Apparatus

Types and Construction of Fire Apparatus

Introduction

Fire apparatus refers to those emergency vehicles involved in fire suppression. Generally, these apparatus are classified by their function, but they can also be classified by construction type. It is important, therefore, that you are able to identify fire apparatus both by function and by construction. The different types of fire apparatus you need to recognize include pumper fire apparatus, mobile water supply fire apparatus, aerial ladder apparatus, elevating platform fire apparatus, and portable pumping units. You also need to be familiar with various types of mounting arrangements and pump drive systems.

In categorizing fire apparatus by function, when we refer to pumping apparatus we include pumpers, mobile water supply apparatus, minipumpers, midipumpers, aircraft fire apparatus, fireboat apparatus, and any number of specialized vehicles that pump water in the battle to suppress a fire. Following the same logic, you can correctly assume that aerial fire apparatus refers to vehicles equipped with aerial ladders and/or elevating platforms that assist in directing the water stream from an elevated position.

All fire apparatus that conform to NFPA standards are classified as standard apparatus; those vehicles that do not meet the applicable standard are considered to be nonstandard. The standard with which you need to be familiar as an EVT is NFPA 1901, *Standard on Automotive Fire Apparatus.*

Pumper Fire Apparatus

The rationale behind the fire department pumper is to provide water at adequate pressure for fire-fighting purposes. This required water stream may be supplied via the pumper from a variety of sources — fire hydrants, apparatus water tanks, and even lakes or ponds.

In addition to the obvious permanently mounted fire pump, the pumper apparatus must also be equipped with intake and discharge pump connections, pump and engine controls, gauges, and other instruments adequate to handle the capacity of the pump being used. Although pumps have been built in just about every imaginable size, current recognized capacities of pumps for modern pumpers are 500 gpm (2,000 L/min), 750 gpm (3,000 L/min), 1,000 gpm (4,000 L/min), 1,250 gpm (5,000 L/min), 1,500 gpm (6,000 L/min), 1,750 gpm (7,000 L/min), and 2,000 gpm (8,000 L/min). The minimum rated discharge capacity for a standard pumper, per NFPA 1901, *Standard on Automotive Fire Apparatus*, is 500 gpm (2,000 L/min).

Pumper fire apparatus are often combined with other fire-fighting tools to increase their usefulness. For example, a triple-combination pumper is equipped with a water tank, hose bed, and fire pump, as well as other accessories, such as extension or roof ladders and forcible entry tools to aid in fire suppression.

In short, pumpers come in a variety of styles and sizes to accommodate a range of fire-fighting needs as illustrated in Figures 2-1 through 2-4. The specific types and models of pumper apparatus are without limit, but we will briefly touch on a few specialized pumpers.

Figure 2-1 Pumper. (Courtesy of Superior Emergency Vehicles)

Figure 2-2 Pumper with enclosed control panel. (Courtesy of Pierce Manufacturing)

Figure 2-3 Pumper. (Courtesy of Superior Emergency Vehicles)

Figure 2-4 Pumper with dual tandem axles. (Courtesy of Superior Emergency Vehicles)

Types and Construction of Fire Apparatus

Pumpers with Foam Capacity

Expressly designed to meet the challenges inherent to flammable liquid fires, some pumpers come equipped with a foam proportioner and a foam delivery device to produce a stream of fire-fighting foam (Figure 2-5). This foam capacity is made possible by either aerating foam nozzles mounted on the roof of the cab of the apparatus or foam proportioning equipment built directly into the main fire pump or by a combination of the two. Because both the nozzle and pump can be operated from within the cab of the apparatus, these pumpers can discharge vast quantities of foam in very little time and achieve quick knockdown without unnecessary exposure of firefighters.

In addition to its special foam nozzles and foam equipment, the pumper with foam capacity also carries the full complement of tools and equipment required by all pumpers. This apparatus adapts to fight other fires in addition to flammable liquid fires.

Figure 2-5 Pumper with foam capability. (Courtesy of Pierce Manufacturing)

Brush/Booster Apparatus

The brush apparatus (also called a booster apparatus) is designed to control ground cover fires. However, it can also be used for small structural fires. This small vehicle is generally a lightweight, easily maneuverable, all-wheel drive apparatus built on a utility-type vehicle chassis to allow access to places larger apparatus cannot reach. Generally, brush apparatus carry a small water tank, very little hose, and a small-capacity pump that may be portable, auxiliary-engine-driven, or power takeoff (PTO) driven. Each of these pumps, whether driven by a separate motor or a power takeoff, gives the apparatus "pump-and-roll" capability whereby it can move and put water on the fire at the same time.

Aircraft Fire Apparatus

Built to carry out a specific task or series of tasks in aircraft crash fire fighting and rescue, the aircraft fire apparatus frequently takes on different characteristics than its cousins in other branches of the fire apparatus family. It often acquires a shape, size, function, and design all its own (Figure 2-6).

Figure 2-6 Aircraft rescue and fire fighting truck equipped with massive foam capability. (Courtesy of Oshkosh Truck Company, Emergency One, Inc.)

The aircraft fire apparatus is designed for a specific use at a specific airport, a use that may or may not be needed at a different airport. The design of each apparatus is strongly impacted by factors such as the size of the airport and the number and types of aircraft that use the airport. All have one thing in common in that they are designed to be fully self-sufficient; they carry an assortment of rescue tools and equipment and are equipped with large-capacity fire pumps. Each carries large quantities of water and foam concentrate and may also be equipped with Halon or dry-chemical extinguishing systems.

It is not unusual for large airports to have several different job-specific units that operate in combination — without conflicting responsibilities. They work together at aircraft fires in much the same way that a pumper and aerial apparatus do at a structure fire; each has its own job to do.

Fireboat Apparatus

The primary responsibility of the fireboat apparatus is to perform water rescue, fire-fighting, and water relay tasks to protect docks, wharves, and boats in waterfront areas. Depending on the required duties, an effective apparatus may be anything from a small, high-speed vessel propelled by water jets to a large, diesel-powered, ocean-going tug or an amphibious vehicle. Regardless of size or function, the boat requires at least one power generator as well as receptacles for receiving shore power, communications, and fresh water. Fireboats are best suited to pumping water through large master stream devices and to supplying additional water for on-shore fire-fighting operations. Each master stream turret on a large fireboat will typically discharge 2,000 to 3,000 gpm (8,000 to 11,500 L/min). Of course, boats have been built to deliver much larger quantities of water.

Initial Attack Fire Apparatus

Initial attack apparatus usually refers to the vehicle that arrives first on the scene. In rural areas especially, the fire department is prone to send a smaller apparatus, equipped with four-wheel drive, that is quick to respond and less expensive to run. Where required, larger apparatus will follow to back up the initial attack apparatus. In urban settings, initial attack fire apparatus could, technically, be any one of several larger vehicles, including pumpers. For our purposes, we classify initial attack fire apparatus as either minipumpers or midipumpers.

Minipumper Minipumpers are small, quick attack pumpers that have been designed for rapid response time and maneuverability (Figure 2-7). Mounted on a pickup chassis, their small size and four-wheel-drive capability allows them to access small spaces and to set up a master stream in areas where a larger pumper might not fit. This apparatus is well suited to controlling smaller fires. The custom-made body of the minipumper incorporates a pump no larger than 500 gpm (2,000 L/min) and may also contain a turret gun that can be fed directly from another pumper.

The minipumper carries the same equipment as full-size pumpers, only in smaller numbers, and may also come equipped with medical supplies that allow it to serve double duty as a rescue unit. Although very reliable for small fires, minipumpers do not have the capability or capacity to be relied upon as attack pumpers, especially at large fires. These units have obvious limitations, but they are extremely useful within these parameters. Since predicting a fire's size at its onset is difficult, it is a good idea to send both a full-size pumper and a minipumper to the fire scene.

Figure 2-7 Minipumper. (Courtesy of Simon Fire Truck Company)

Midipumper Like the minipumper, the midipumper is well suited to service calls such as dumpster fires and grass fires that do not require the capacity and personnel of a full-size pumper (Figure 2-8). Unlike its smaller cousin, however, it can launch an initial attack on a larger fire. The midipumper is quite commonly utilized as the initial attack fire apparatus at most fire scenes in rural areas, with a pumper backing it up where required. In urban areas, the midipumper serves as an initial attack apparatus for grass fires and smaller, contained fires.

The most significant differences between the minipumper and the midipumper are size, pump capacity, and amount of equipment on board. The midipumper tends to be larger than the minipumper and smaller than the regular pumper, yet carries the same type of equipment as the full-size pumper. Built on a chassis of over 12,000 pounds (5,443 kg) gross vehicle weight (GVW), the midipumper supports pumps as large as 1,000 gpm (4,000 L/min) and carries the same hose, ground ladders, and equipment as its full-size counterparts. It is not uncommon for the midipumper to carry medical equipment and pull double-duty as a rescue vehicle.

Figure 2-8 Midipumpers. (a) The pump panel on the side is the traditional North American method of manufacture. (b) More pumpers are now being manufactured with the pump panel on the rear of the vehicle as this is a much safer location. (Both: Courtesy of Superior Emergency Vehicles)

Mobile Water Supply Fire Apparatus

The purpose of the mobile water supply fire apparatus is to transport water. Also commonly referred to as a tanker or tender, this apparatus has a water tank with a minimum 1,000 gallon (4,000 L) capacity and includes a pump as well as limited hose body capacity (Figure 2-9). Mobile water supply apparatus are designed to supplement the limited water supply of standard pumpers. Even in urban areas where water is readily accessible, these tankers can be extremely beneficial during an extended fire-fighting operation. They can act as a reservoir, can be used as support vehicles, can carry water to areas beyond a water system or to areas where the water supply is inadequate, and can participate in shuttle operations. The tanker's support role in fire suppression is an important one.

Figure 2-9 Tanker. (Courtesy of Pierce Manufacturing)

Each fire department has restrictions as well as requirements that will dictate the size of the water tank on its mobile water supply fire apparatus. Geographical factors such as local terrain and bridge weight limits factor into determining the capacity of the apparatus's tank. Monetary constraints will also play a role. Additionally, all tankers in the fire department or responding area should be of similar size to facilitate smooth flowing shuttles.

Regardless of the tank size, all mobile water supply apparatus must satisfy certain construction requirements if the water is to be transported quickly and without incident. Single-rear-axle vehicles should be limited to a tank capacity of 1,500 gallons (6,000 L) or less. When a larger capacity tank is required, the apparatus should be built on either tandem rear axles or a tractor-trailer chassis. Additionally, the apparatus's components must be adequate to support the size and weight of the water tank. It is crucial that the tank is supplemented with proportionate apparatus components, such as:

- Sufficient suspension and steering
- Suitably sized chassis
- Appropriately sized engine for tank size and local terrain
- Adequate braking ability
- Accurate tank mounting
- Correct and reliable tank baffling

For more information about the standards for mobile water supply apparatus, refer to NFPA 1901, *Standard on Automotive Fire Apparatus.*

Aerial Apparatus

Aerial fire apparatus are divided into two categories: aerial ladders and elevating platforms.

Aerial Ladder Apparatus

The purposes of the aerial ladder apparatus are to accomplish rescue and ventilation operations, apply an elevated master stream, and gain access to the upper levels of structures. This vehicle consists of a special truck chassis that has been equipped with a permanently mounted power-operated aerial ladder (Figure 2-10). It also is equipped with the regular complement of ground ladders, tools, and other fire-fighting equipment.

Figure 2-10 Aerial ladder truck. (Courtesy of Pierce Manufacturing)

The ladder, which is operated with hydraulic pumps, cylinders, and motors, has a working height of 50 feet to 135 feet (15 m to 41 m), measured from the ground to the highest ladder rung. It is designed and built to perform safely under all potential fire and rescue conditions. As an added safety precaution, the truck must have an electric or mechanical backup system for the ladder if it is to conform to the specifications of NFPA 1901.

NFPA 1901 also stipulates that all ladders be comprised of at least two sections and be constructed of metal, typically heat-treated aluminum alloy or steel. The ladder's sides usually are trusses assembled to form triangles, thus allowing tension and compression stresses to be spread over the entire length of the ladder. This truss construction also provides the strength necessary for the ladder to withstand the stress of the job, regardless of whether the tip is supported or unsupported.

Aerial ladders may be mounted on either a two- or three-axle, single-chassis vehicle equipped with dual rear wheels or on a three-axle tractor-trailer vehicle that has steerable rear wheels on the trailer (Figure 2-11). Although a tiller operator is required to steer the rear wheels of the tractor-trailer vehicle, it is easier to guide through narrow streets and heavy traffic than the single-chassis vehicle.

Figure 2-11 Aerial tiller. (Courtesy of Emergency One, Inc.)

Elevating Platform Fire Apparatus

Sometimes our most useful pieces of equipment become even more helpful in certain situations with slight modification. The elevating platform fire apparatus is a case in point. In the simplest of terms, an elevating platform fire apparatus is an aerial ladder apparatus that has been permanently equipped with a passenger-carrying or work platform at the uppermost boom (Figure 2-12).

Aerial elevating platform apparatus combine the convenience of a safe, easy-to-climb aerial ladder with the benefits of a safe work area. They are always single-chassis vehicles and usually have three axles. On most elevating

Figure 2-12 Aerial elevating platform apparatus. (Courtesy of Pierce Manufacturing)

Figure 2-13 Telescoping aerial platform apparatus. (Courtesy of Superior Emergency Vehicles)

platform apparatus, the aerial device is mounted at the rear of the vehicle, but some models offer midship mounting.

The construction of the aerial boom can be articulating, telescoping, or a combination of the two. Articulating booms consist of two sections that are joined by an elbow-like hinge and fold onto one another when not in use. Telescoping devices have two or more sections that slide into one another for storage and are made either of box-beam or tubular truss-beam construction (Figures 2-13, 2-14, and 2-15). One huge advantage of the telescoping elevating aerial apparatus is its extensive reach capability, even when the master stream is being operated.

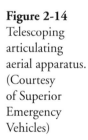

Figure 2-14 Telescoping articulating aerial apparatus. (Courtesy of Superior Emergency Vehicles)

Figure 2-15 Telescoping aerial platform apparatus. (Courtesy of Emergency One, Inc.)

Other than in operation of the aerial device, there is no difference between the telescoping and articulating aerial platform apparatus. Both are generally powered by hydraulics and require the same fire-fighting equipment. NFPA 1901 requires all platform apparatus to be equipped with a permanently mounted turret nozzle that is supplied by a water system built into the booms, as well as with electrical, air, and hydraulic outlets to facilitate fire-fighting and rescue operations. The apparatus must be equipped with two operating control stations — one at street level and one in the platform — with a communication system between the two. For obvious reasons, the apparatus must also have a backup to the hydraulic system.

The design of the aerial ladder platform must consider safety requirements. To meet NFPA 1901 the platform must be constructed of metal. Other safety requirements include:

- A leveling system is essential to ensure that the platform is always horizontal.

- The floor area of the platform must be a minimum of 14 square feet (1.3 m^2).

- The platform must be completely enclosed by a railing that has no bottom opening greater that 24 in (610 mm) in height.

- Each platform should include two gates below the top railing to allow access and exit.

- A 4-inch (100 mm) high kick plate is required at floor level to prevent people from slipping off the platform.

- The kick plate should have a drain opening measuring at least ¼ in (7 mm) to prevent water buildup.

- The platform should be equipped with a heat protective shield to protect occupants from radiated heat.

Fire Department Portable Pumping Units

The fire service utilizes stand-alone engines to power fire pumps in a variety of specialized applications, including brush and airport crash trucks, but the most common application is as a portable pump (Figure 2-16). This stand-alone capability arises from the fact that the pump engine operates independently of the drive system. This allows the unit to be mounted on a skid, trailer, or apparatus, as dictated by the needs of the fire department. The portable pumping unit offers a great deal of flexibility. The separate engine not only allows the unit to be mounted or transported anywhere it will fit but also makes it ideal for pump-and-roll operations where it is advantageous to pump water while the vehicle is still in motion.

Capacity for the typical portable pump is limited to 500 gpm (2,000 L/min). Although impractical for use as a primary fire pump, it adapts

Figure 2-16 Portable pumping units.

well when used within its limitations. The pump is versatile, portable, simple to maintain, and less expensive to build and purchase than most other fire pumps.

Major Types of Fire Apparatus Construction

Fire apparatus are designed from the bottom up with a broad spectrum of features to meet the specific needs of different fire-fighting applications. Each fire department has different requirements for its equipment, depending on the climate, surrounding terrain, neighboring industry, and the potential for different types of fire. The wear and tear the apparatus will receive, combined with its application, establishes the characteristics essential to the fire apparatus. Manufacturers have developed a variety of mounting arrangements and pump drive systems in order to meet these requirements.

Cross-Mounted Engine Drive

This economical multipurpose apparatus meets both UL and ISO certification as a Class A pumper. It is ideal in settings that require both the power

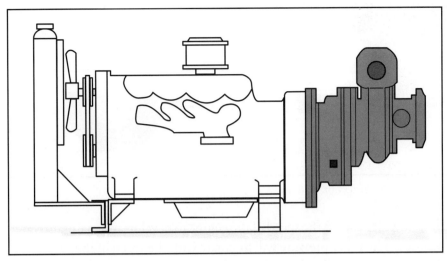

Figure 2-17 Auxiliary-engine-driven pump. (Courtesy of Waterous Company)

to fight structural fires and the pump-and-roll capability to combat ground cover fires. Mounted midship on the apparatus, the auxiliary-engine-driven pump draws no power from the vehicle's main engine (Figure 2-17). The pumps, rated at from 500 gpm to 750 gpm (2,000 L/min to 3,000 L/min), contain enough power to extinguish structural fires and can run while the apparatus is in motion.

Power Take-Off Drive

The power take-off (PTO) drive allows pump-and-roll operations, but they are not as effective as those of the portable pump unit. If the apparatus is designed for pump-and-roll, ensure that a pressure gauge is mounted inside the cab; the operator should drive by the pressure gauge instead of the speedometer during these maneuvers.

The most common applications of the power take-off unit include brush trucks, minipumpers, and mobile water supply apparatus. However, as modern technology continues to develop full-torque power take-offs that support higher-capacity pumps, larger pumpers will undoubtedly integrate the full-torque PTO unit to incorporate the flexibility of pump-and-roll operation.

Conventional PTO units limit the capacity of the fire pump to 500 gpm (2,000 L/min); this is mostly because the PTO unit is mounted on the transmission, whose housing can withstand only limited horsepower (70 hp [52 kW]).

The PTO unit is mounted in the transmission of the apparatus through a universal-joint shaft (Figure 2-18). To ensure dependable, smooth operation, the pump's gear case must be mounted high enough above the chassis to avoid damage but with a minimum of angles in the drive shaft. An idler gear in the transmission of the apparatus runs the power take-off unit that drives the pump. Powered by the transmission, the PTO unit and therefore

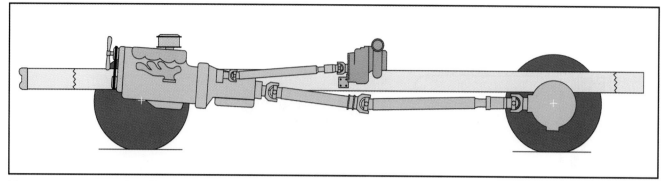

Figure 2-18 A power take-off drive train arrangement. (Courtesy of Waterous Company)

the pump depend on the speed of the engine to develop and maintain water pressure. The water pressure will fluctuate with the speed of the vehicle. At the same time, the clutch maintains some control over the pump; when it is disengaged to shift gears or stop the vehicle, the pump will stop.

Front-Mount Pumps

Apparatus ranging from tankers to rural pumpers and some urban pumpers often come equipped with a front-mount pump. The frame rails of the apparatus extend beyond the front bumper, and the pump nestles between the radiator and the bumper on a support between the two rails (Figures 2-19 and 2-20). This mounting system not only provides easy access to the pump for operation and maintenance but also protects the pump in the event of a front-end collision. Nevertheless, it does have its disadvantages.

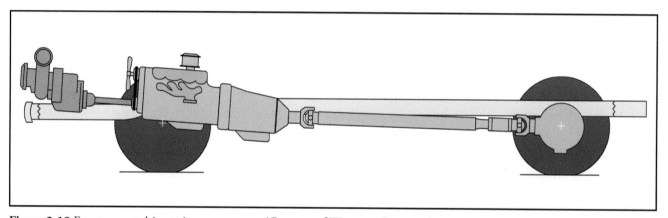

Figure 2-19 Front-mount drive train arrangement. (Courtesy of Waterous Company)

Figure 2-20 Front-mount pump on a fire truck. (Courtesy of Superior Emergency Vehicles)

Types and Construction of Fire Apparatus

The external lines of the pump and gauges are prone to freezing in cold climates, and improper installation can make the pump vulnerable in the event of a motor vehicle collision.

Nothing about the installation or instrumentation of a front-mount pump is standardized. The pump's capacity may be 500 gpm (2,000 L/min), 750 gpm (3,000 L/min), 1,000 gpm (4,000 L/min), or 1,250 gpm (5,000 L/min), depending on the limitations of the engine that powers it. Each capacity pump differs in design. Additionally, the available gear ratios vary according to the operating characteristics of the engine and the desired capacity of the pump (usually between 1½:1 and 2½:1).

The units do have some features in common. A power take-off (PTO) unit fastened to the front of the engine's crankshaft powers the front-mount pump. Friction and a positive-drive clutch transmit power to the pump through a gearbox and to a clutch connected by universal-joint shafts to the front of the engine fan pulley (crankshaft). The gearbox uses a step-up gear ratio that causes the pump's impeller to rotate more rapidly than the engine.

The front-mount pump is engaged and controlled from the actual pump location. It can be used for pump-and-roll operations because it runs independently of all drive-system components except the engine; however, the operator should drive the apparatus by a pump-pressure gauge rather than by the speedometer for these maneuvers.

Midship Transfer Drive

The majority of pumpers have the fire pump mounted laterally across the frame of the apparatus, behind the engine, using a split drive shaft arrangement (Figure 2-21). In one variation of the midship transfer drive, the pump is mounted at the rear of the apparatus (Figure 2-22). Midship transfer drive units offer more flexibility in capacity and design because they are not limited to the small space available for front-mount installation. They also provide higher rated flow because they use all of the engine's power to drive the pump.

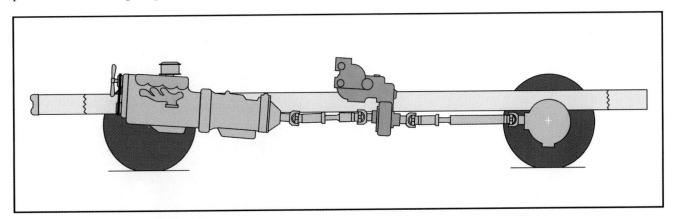

Figure 2-21 Midship transfer drive. (Courtesy of Waterous Company)

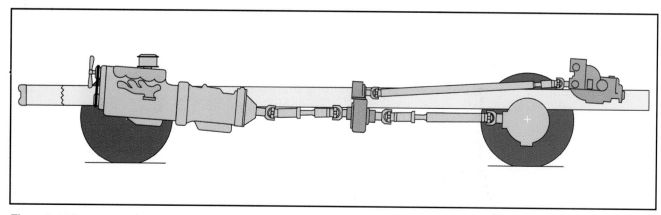

Figure 2-22 Rear-mount drive: a variation of the midship transfer drive. (Courtesy of Waterous Company)

Most pumps operate over a range of capacities. The maximum capacity that the midship transfer drive can obtain depends on the impeller, piping, gear ratio, engine horsepower, and the size of the pump. A double-suction impeller can supply as much as 2,000 gpm (7,570 L/min).

A split shaft arrangement with a pump-drive transmission powers the unit. The pump transfer case is mounted between the transmission and the rear axle of the vehicle and is controlled from the cab by a power shift arrangement that may be mechanical, electrical, hydraulic, or air-operated. A simple shift of a gear diverts power from the rear axles to the fire pump, which is then run by either a series of gears or a drive chain (Figure 2-23). The midship transfer drive is sometimes subdivided to include single-stage

Figure 2-23 Cutaway showing construction of chain drive for a fire pump. (Courtesy of Waterous Company)

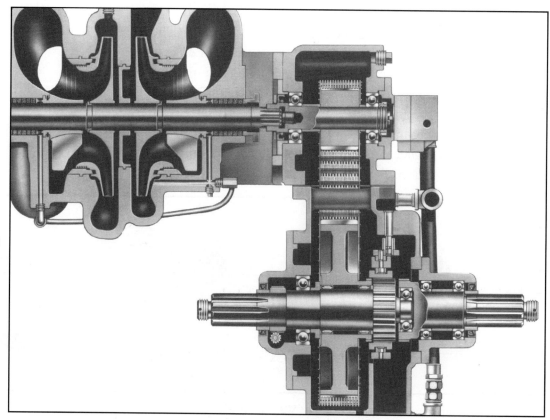

Types and Construction of Fire Apparatus

midship and two-stage midship drives. Whereas the single-stage midship utilizes one impeller, the two-stage midship incorporates two separate impellers, one for capacity and one for pressure.

The road transmission must be in the correct gear if the pump is to operate at peak efficiency. Pump-and-roll operations are *not* possible because using the pump requires transferring the power from the rear axle to the pump drive gears.

Summary

While this chapter has covered fire apparatus whose primary function is either aerial or pump related, it is important to remember that foam and chemical units, squad and rescue apparatus, and utility vehicles may also be equipped with fire pumps.

Fire apparatus come equipped with a variety of features to suit almost every application. Variations among apparatus include the size of the chassis; type and capacity of the pump; presence or lack of aerial devices; performance in acceleration, speed, turning radius, drive-wheel horsepower, and so forth; quantity and type of equipment; and water tank capacity.

These characteristics allow fire departments the freedom to pick and choose the proper apparatus for their needs and, where resources allow, for each fire. The benefits — time, money, and lives saved — realized from the ability to send either job-specific or multipurpose apparatus to each fire can be enormous.

Chapter 2 Review Questions

True or False

Write *True* or *False* before each of the following statements. Correct those statements that are false.

_____ 1. Pumping apparatus that conform to NFPA 1901 are classified as nonstandard.

_____

_____

_____ 2. The standard for the design of aerial apparatus is NFPA 1924.

_____

_____

_____ 3. Aerial ladders are under stress when the tip is supported.

_____

_____

_____ 4. The minimum required floor area of an aerial platform is 1.3 m².

_____

_____

_____ 5. Two control stations are required on all aerial apparatus.

_____

_____

_____ 6. The minimum rated discharge capacity for a standard pumper is 500 gpm (2,000 L/min).

_____

_____

Short Answer

Write your answers to the following questions on the blank lines provided.

1. According to NFPA 1901 what is the minimum rated discharge capacity for a standard pumper?

_____

2. To be considered a triple-combination pumper, what equipment must the truck have?

_____

3. What is the main difference between minipumpers and midipumpers?

_____

4. A truck's maximum weight is known as what?

_____

5. Large water tankers are often used to deliver water to the fire ground. What term is used to describe this operation?

_____

6. List the two main types of aerial apparatus.

_____

_____

7. List a minimum of three ways that power is transferred from a vehicle engine to the pump.

_____

_____

_____

8. Which of the pump mounting and drive arrangements draws no power from the main engine of the apparatus?

_____

9. The conventional power take-off unit is most commonly used for what fire-fighting applications?

_____

_____

Multiple Choice

_____ 1. Which of the following is *not* a disadvantage to the front-mount pump?

    a. It has pump-and-roll capability.

    b. The lines are prone to freezing in cold climates.

    c. It is vulnerable in front-end collisions.

    d. All of the above.

_____ 2. Which of the following does *not* have pump-and-roll capability?

    a. Front-mount pump

    b. Auxiliary-engine-driven pump

    c. Midship transfer drive

    d. Cross-mounted engine drive

Chapter 3
Fire Pump Theory and Maintenance

Fire Pump Theory and Maintenance

Introduction

A fire pump can be depended upon to work in an emergency only if it is properly operated and maintained. To maintain fire pumps properly, it is crucial that emergency vehicle technicians understand the basic operating principles of pumps, the different ways to drive and control them, and the basic maintenance procedures necessary to keep them in peak operating condition. This chapter will highlight the basic physics of fire pump operation, describe the operating principles of positive displacement and centrifugal pumps, and outline the routine and periodic maintenance necessary to ensure peak operating performance of those fire pumps.

Purpose of the Fire Pump

Fire pumps enhance the water supply pressure available from hydrant systems, water tanks, and static sources to produce effective water streams for fire-fighting purposes.

As our water systems have evolved over the decades, so have our means of delivering effective water streams. Early fire-fighting efforts consisted of bucket brigades that transported water from the nearest source in buckets and threw it on the fire. Portable tubs with hand-operated pumps made way for larger tanks mounted on horse-drawn apparatus with double-action hand pumps, but both water tanks were still supplied by bucket brigades.

Hand-operated pumping apparatus were eventually equipped with a leather hose able to draft water from nearby sources, but this process was limited to areas with a handy source of water. Underground water pipes, combined with the technology that first brought mill pumps and then displacement (or rotary) pumps followed by reciprocating steam pumps and centrifugal pumps have shaped fire-fighting equipment and simplified the problem of water transportation.

Today, the centrifugal fire pump is accepted as the standard fire pump. It has made earlier types of pumps obsolete, although not extinct.

Physical Characteristics of Water

Without an understanding of water's basic physical characteristics, firefighters would still be frantically throwing buckets of water in an effort to do their jobs, and emergency services technicians would have no pumps to maintain. Advances in fire-fighting equipment have mirrored the advances of science and of the technology for using water from distribution systems.

The actual physical properties of water determine the way it is used in fire suppression applications. This means that fire pump theory revolves around the principles of the water that is pumped through the pumping system. Thus, we need to revisit your high-school science lessons to refresh your knowledge of the principles of water movement and pressure.

Properties of Water and Water Movement

Water has many characteristics that make it an effective fire-fighting tool. Scientifically speaking, water is not easily broken down by its environment; as a result, it is a very effective coolant. Water (H_2O) is comprised of three atoms — two of hydrogen and one of oxygen. In its natural state, each of these elements is a gas, but once they have combined to form a molecule of water, separating them is very difficult, regardless of outside influences. A water molecule remains a water molecule in liquid, gas, or solid form.

On a more general level, water has no shape of its own, but it does have measurable volume. It has mass and is heavier than gas. Water assumes the shape of its container, has little to no adhesive quality, and cannot be pulled or lifted. It can, however, be pushed from one location to another when pressure is applied at the same time the water is restrained to direct the flow.

While at rest, drops of water in a container sit on top of one another with no physical adhesion. The pressure applied against one drop is shared with every other drop it touches without decreasing. This pressure is transmitted equally in all directions.

Where the drops contact a restraining surface, the pressure is applied at right angles to the surface. At the same time, the downward pressure of the water is proportional to its depth; as long as the level of water in separate containers is the same, the pressure will be the same, regardless of the container's shape.

Water is virtually impossible to compress and therefore cannot store pressure. In its liquid state, a pressure of 33,000 psi (231,000 kPa) decreases the volume of water by only one percent. When that pressure is removed, there is no significant expansion of water, and as a result the pressure is relieved.

It is this characteristic that allows water to be pumped so efficiently. Water moves easily and quickly into areas like the hose or nozzle where the pressure is lower because it cannot compress when pressure is applied. Add to this the fact that water in motion tends to remain in motion without changing direction and you explain the phenomenon of water hammer—hoselines shut down suddenly cause a high-velocity "crash" at the end of the system.

Water in motion is governed by several important characteristics:

- The size of the nozzle opening and the velocity of the water determine the amount of water that flows through a hose line.

- When water moves through the line, the friction of the water against the surface of the waterway creates a resistance to its movement that is referred to as friction loss.

- Water loses its pressure while in motion, due to friction loss in the pipe.

- Gravity can cause pressure differences as the result of changes in elevation between different points in a water supply system.

- Water in motion tends to remain in motion with no change of direction.

Water Pressure

Water pressure results when water is pushed into a restrained area more quickly than it can get out. This pressure may be generated naturally, through gravity and the weight of the water; mechanically, by way of a pump; chemically, through a mixture of substances; or by any combination of the three. In fire-fighting terms, pressure is defined as the velocity of water in a pipe or hose. For our purposes, pressure is measured in pounds per square inch (psi) or kilopascals (kPa). The basic types of water flow pressure are:

- Atmospheric pressure

- Head pressure

- Static pressure

- Normal operating pressure

- Residual pressure

- Flow pressure

Atmospheric pressure, the force the earth's atmosphere exerts on everything on the planet, is measured through a barometer by comparing the weight of the atmosphere with that of a column of mercury. The height to which the mercury climbs is in direct relation to increases in atmospheric pressure.

This atmospheric pressure will push water up the suction hose to the eye of the impeller in the pump. (Contrary to popular belief, pumps do not suck the water into themselves; the atmosphere pushes the water in.)

Head distance (or head height), the distance the water supply sits above the discharge, is translated into *head pressure* through a simple equation. To determine head pressure, divide the distance in feet between the water source and discharge by 2.304 inches (or in meters, by 0.1), the distance that 1 psi (7 kPa) will raise a column of water.

When the water is not moving, the exerted pressure is said to be *static pressure*. By water flow definition, static pressure is the probable energy stored and available to push the water through the pipe, fittings, fire hose, and adapters crucial to fire fighting. In reality, however, municipal water systems are seldom, if ever, motionless. Static pressure usually refers to the normal pressure existing in a water supply system before a flow hydrant is opened. It may be influenced by atmospheric pressure or by natural or electrical forces.

The flow of water through a distribution system varies depending on the time of day. *Normal operating pressure* is the pressure that occurs when water is flowing. It is a calculation of the average of the total amount of water consumed each day over a one-year period; it is neither the pressure during peak consumption periods nor during slower times but an average of both.

The pressure remaining at a specific location in a water supply system when the water is flowing is referred to as *residual pressure*. It is that portion of the full amount of available pressure that is not used to overcome friction or gravity while propelling water through the pipe, fittings, fire hose, and adapters. Residual pressure is generally measured at one hydrant while water is flowing from another hydrant.

Flow pressure is the forward velocity force or pressure of flowing water through a discharge opening such as the hydrant discharge or the end of a nozzle. The velocity of the water stream being propelled forward through the system and fire-fighting equipment exerts this flow pressure and can be measured with a pitot tube and gauge.

Fire Pump Theory

The majority of water hydrant systems do not have adequate pressure to support effective fire fighting. As a result, firefighters have incorporated the principles of water and electrical pressure in the form of fire pumps to produce effective water streams for fire suppression.

Early water pumps included arm-powered and steam-powered mechanisms. On a simplified level, everything from the bucket brigade and reciprocating steam pump to today's priming pumps is typically a *positive displacement pump.* These require positive action, usually physical, to force a specific amount of water from the pump body with each operating cycle. The *standard centrifugal fire pump,* on the other hand, depends on the velocity of the water produced by centrifugal force, rather than on a positive action, to provide the necessary pump discharge flow. Its compact design, dependability, hydraulic features, and easy maintenance, combined with the range of possible drivers — electric motors, steam turbines, and internal combustion engines — have made it a favorite in the fire-fighting industry. All pumps in use today are either positive displacement or centrifugal.

Pressure vs. Flow

It is very important for the emergency vehicle technician to understand the difference between pressure and flow.

1. Pumps make flow; they do not, and cannot, make pressure.

2. Resistance to the pump's flow causes pressure.

3. The faster the pump turns, the more flow it will create for the same sized nozzle. A higher pressure in the line will also result. This has led people to think that the pump made the higher pressure when, in fact, if you were to open a few more nozzles and keep the engine speed the same, the system pressure would drop. The nozzle, and therefore the resistance to the water flow, causes the system pressure.

4. All modern pumps have a rating that requires them to produce a given amount of flow (liters or gallons) against a given amount of pressure. As a pump wears out, its ability to produce its rated flow against the stipulated pressure will diminish, and the pump will have to run at a faster (higher) engine speed to produce the same pressure. The section on pump testing later in this chapter explains the importance for a pump to produce its rated flow against a given pressure at a specified engine speed (rpm).

Positive Displacement Pumps

Because fire pumps are governed by the principles of water, they must incorporate the basics of hydraulic law into their operation. Both piston- and rotary-driven positive displacement pumps depend on the near incompressibility of water. They work on the premise that when pressure is applied to a confined liquid, all of that pressure is spread at the same level and in all directions throughout the liquid.

The *piston pump* contains a piston that slides back and forth within a cylinder and generates flow that in turn operates the intake and discharge valves. When the pump begins operating, the cylinder contains only air. As the piston moves forward, it compresses the air inside the cylinder, producing a higher pressure inside the pump than the atmospheric pressure in the discharge manifold. This increased pressure triggers the discharge valve, and air escapes through the opening. Once the piston completes its forward stroke and stops, the pressures equalize and the discharge valve closes.

When the piston starts its return stroke, the space behind the piston increases, and the pressure in the cylinder decreases. This creates a partial vacuum that opens the intake valve and draws in air from the suction hose (Figure 3-1). The migration of this air lowers the pressure in both the hose and the intake area. Atmospheric pressure then pushes the water into (or up the hose) until the piston completes its stroke and the intake valve closes. Repeated strokes of the piston evacuate the air out through the discharge until all of the air in the cylinder has been replaced with water. The pump is now primed (Figure 3-2).

Additional strokes of the piston will now force water out the discharge valve and through the pump. The forward strokes discharge water, and the return strokes fill the pump.

The piston pump has not been used as a major pump on pumping apparatus for many years, mostly because of its pulsating water stream (no water flow on the return stroke) and its susceptibility to wear. However, some are still used in small capacity, high-pressure fire-fighting applications and as priming pumps during drafting operations.

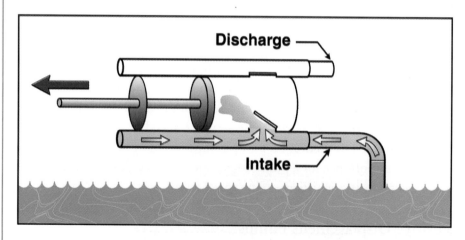

Figure 3-1 The partial vacuum created as the piston begins the return stroke causes the intake valve to open. This allows air from the suction hose to enter the pump.

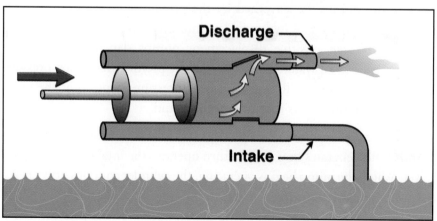

Figure 3-2 Once all the air has been evacuated, only water will be pushed through the pump.

The *rotary pump* also has been delegated for the most part to the role of booster pump or priming pump. From a design standpoint, the rotary pump is the simplest of the apparatus fire pumps. Although there are many different designs (mainly in the shape of the rotors) for this pump, they all operate on the same principle.

The *rotary gear pump* consists of two intermeshing rotors within a close-fitting casing (Figure 3-3). In operation, the rotors turn in opposite directions so that the teeth move away from each other on the suction side and toward each other on the discharge side. This movement traps air or water from the suction chamber in pockets between adjacent teeth and the casing and then carries it toward the discharge chamber. As each tooth leaves the casing wall, it releases the trapped air or water into the discharge chamber. The meshing of the teeth as they return toward the suction side squeezes the air or water out of the chamber into the discharge manifold. At the same time this prevents the return of air or water to the suction side. Rotary pumps, partially because of this action, are called positive-displacement pumps. They are sometimes termed self-priming

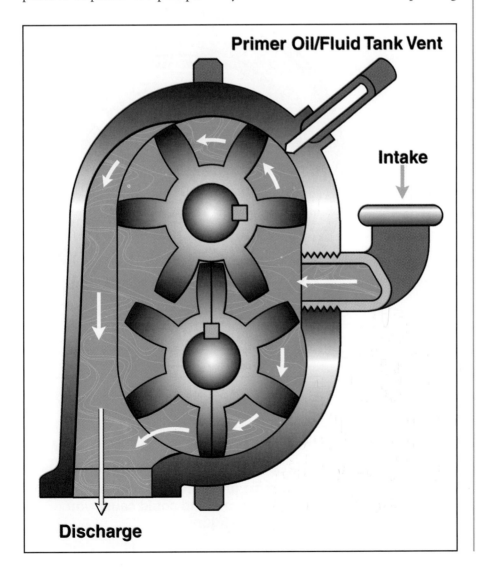

Figure 3-3 Rotary gear priming pump.

pumps because they can expel air as well as water and therefore can prime themselves. Like piston pumps, rotary gear pumps tend to wear down through normal use.

The *rotary vane pump* is designed to automatically compensate for wear. A rotor is mounted off-center inside the housing, with a larger space between the rotor and housing at the intake than at the discharge (Figure 3-4). When the rotor starts spinning, its vanes trap air against the casing. As the vanes rotate, this pocket gets smaller, compresses the air, and increases the pressure within. When the pressure reaches its maximum level, the trapped air is forced out of the pump, thus lowering the pressure. Atmospheric pressure then draws water into the system. At this point the pump is primed and will force water out in much the same way that it forced out the air.

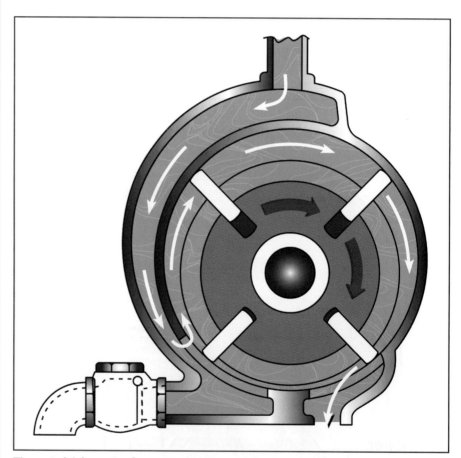

Figure 3-4 Schematic of a rotary vane pump.

Caution:
Never completely close off the discharge from a positive discharge pump; the extreme pressure that can be instantly created will destroy the pump and could cause you personal injury.

Centrifugal Pumps

Today, positive action pumps have made way for centrifugal force. With increased production of high-speed engines throughout the automotive industry, centrifugal pumps are probably more widely used than all other types. They are the main pump on most modern apparatus because they are reliable, compact, easy to maintain, can handle more contamination in the water, and are not restricted to pumping a specific amount of water with each revolution.

The centrifugal pump operates on the premise that you can impart velocity to water and convert it to pressure within the pump itself. Simply put, it converts kinetic energy to velocity and pressure energy. A spinning disk produces velocity in water by throwing it toward the outer rim of the disk and converts it to pressure by confining the water within the container.

The two main components of a centrifugal pump are an impeller and the case in which the impeller rotates (Figure 3-5). The impeller is comprised of a pair of rotating discs called shrouds, separated by curved partitions called vanes. It is mounted off-center within the casing to create a water passage called the volute. As the impeller revolves, water introduced through the suction tube enters the impeller at the center through the "eye" (Figure 3-6). It is then picked up by the curved vanes as they revolve with the impeller, thrown to the outer edge by centrifugal force, and into the open space of the casing.

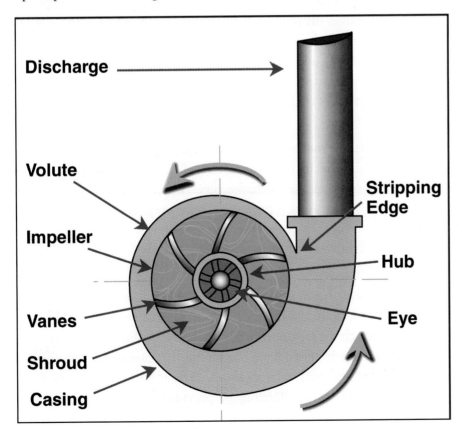

Figure 3-5 The major parts of a centrifugal pump.

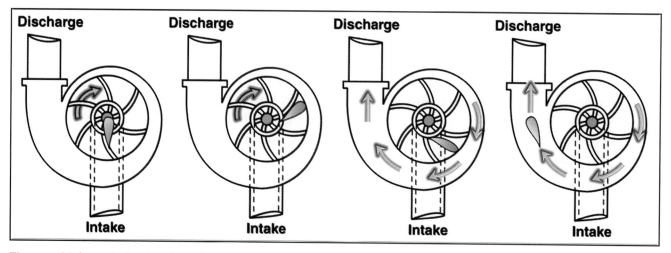

Figure 3-6 Schematic showing the path a single drop of water takes through a centrifugal pump.

Since the circumference of the impeller is greater at the outer edge of the vanes than at the eye, the outer edge of the impeller travels faster than the surfaces at the eye. Thus, the water's velocity increases as it passes through the eye and travels through the impeller to be thrown off into the casing. As the rotations of the impeller increase, the velocity with which the water is thrown off also increases.

The cross-sectional area between the outer edge of the impeller and the wall of the casing, called the volute, constantly increases toward the discharge outlet. Its increasing size is necessary because water is thrown from around the entire circumference of the impeller, and hence the total quantity of water increases greatly toward the discharge outlet. The design of the volute enables the pump to handle this increasingly greater quantity of water and permits the velocity of the water either to remain constant or to decrease gradually and thus maintain the required continuity of flow. Some pump designs direct the flow of water toward the discharge outlet by means of a series of stationary diffusion vanes fastened to the inner wall of the pump casing.

In principle, the movement of the impeller simply creates a velocity in the water, and the velocity in turn is converted into pressure as it approaches the confining space of the discharge pipe. Water under pressure on the discharge side of the casing is prevented from flowing back into the pump by the close fit between the casing and impeller at the entrance to the suction inlet, by the speed of the impeller, and by the equal pressure in the pump casing.

Since the centrifugal pump is not of the displacement type, it has no valves or other blockades within the pump proper and presents a continuous waterway through the pump from the intake to the discharge outlet. This type of pump closely resembles a blower fan that cannot exhaust enough air to create the necessary vacuum to prime itself. Other means involving mechanical devices must therefore be provided for priming centrifugal fire pumps.

The terms *single-stage, two-stage, multiple-stage,* and *duplex-multistage* are often used when dealing with centrifugal pumps. We may say that each impeller is a stage. A single-stage pump, therefore, has one impeller (Figure 3-7). Likewise, a two-stage pump has two impellers. A pump comprising two or more single-stage pumps connected by a common drive is called a multistage centrifugal pump (Figure 3-8).

Multiple-stage pumps can be arranged in series or parallel. In series pumps, water flows from one intake, through one stage, into another intake, through another stage, and so on through as many stages as the pump may have, and finally, out a single discharge. In parallel pumps, all the stages take in water from a common manifold and discharge it into another common manifold. Valves between the stages can make a pump capable of both parallel and series operation.

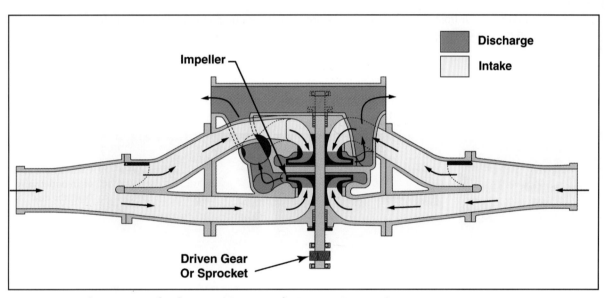

Figure 3-7 Single-stage centrifugal pump. (Courtesy of Waterous Company)

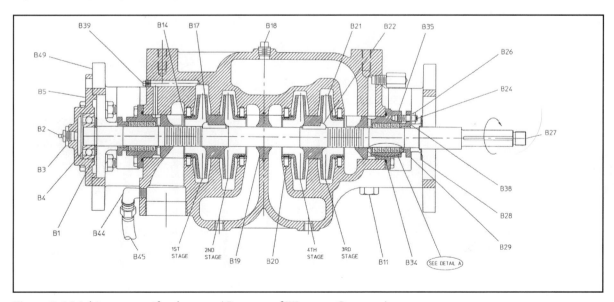

Figure 3-8 Multistage centrifugal pump. (Courtesy of Waterous Company)

Valves for Fire Pumps

Many valves are used on fire apparatus to control water flow. They include intake valves, intake relief valves, transfer valves, relief valves, and outlet valves.

INTAKE VALVES

Any intake or outlet valve on the pump that is larger than 3 inches (76 mm) must use a gated valve of the slow-operating design; that is, it cannot open or close faster than three seconds (**Figure 3-9**). This valve also must not take longer than 10 seconds to open fully.

INTAKE RELIEF VALVES

An intake valve of 4 inches (100 mm) or larger must be equipped with an intake relief valve. The automatic adjustable valve will limit the incoming water pressure and ensure dangerous water pressure spikes (water hammer) are reduced (see Appendix K for more information on relief valve operation).

TRANSFER VALVES

When using dual-stage pumps, the operator must be able to switch from single stage (parallel flow) to dual stage (series flow) while the pump is under pressure. This is done with a transfer valve (**Figure 3-10**). The transfer valve must have a manual override. Common methods of activation are electrical, vacuum, or air pressure.

RELIEF VALVES

Discharge relief valves dump water from the high-pressure outlet side of the pump to the intake side of the pump (**Figure 3-11**). They control maximum system pressure only. An indicator on the pump panel must be illuminated when the valve is in operation.

OUTLET VALVES

Every outlet must be equipped with a discharge valve that can be opened or closed at any operating pressure (**Figures 3-12 and 3-13**). The operator must be able to lock the discharge valve in position so that it cannot be inadvertently changed during operation (**Figures 3-14 and 3-15**).

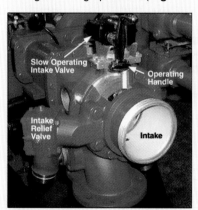

Figure 3-11 Cutaway of a discharge relief valve and the electrical connection.

Figure 3-12 Cutaway picture of an outlet valve. Note the dual control rods, which allow the valve to be operated from either side of the truck.

Figure 3-9 Cutaway view of both an electrically operated slow-operating intake valve and the intake relief dump valve. Note the operating handle for the slow-operating valves; this handle would be on the outside of the panel.

Figure 3-13 Large air-operated valve.

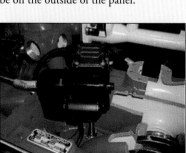

Figure 3-10 Cutaway of an electrical transfer valve. Note the shaft for manual operation on the left.

Figure 3-14 Locking intake valve. A quarter turn to the right locks the valve in position. Note the electrical switch to light an indicator on the pump panel if the valve is open.

Figure 3-15 Lock valves. Note the locking mechanisms on the ends of the valve handles.

Fire Pump Theory and Maintenance

The mobile multistage pumps in most common use by the North American fire service today are two-stage parallel-series pumps (Figure 3-16). This type of pump is a multistage pump connected with "waterways" so that a transfer valve (also called a changeover valve) can either allow each pump to take suction and discharge independently or direct the discharge from one pump into the intake of the second. In parallel operation the discharge is equal to the sum of the capacities of the pumps at the pressure for which each pump is designed. For example, with each pump operating at a capacity of 264 gpm (1000 L/min) at 145 psi (1000 kPa), the total discharge will be 528 gpm (2000 L/min) at 145 psi (1000 kPa). In series operation, the discharge is equal to the capacity of the first pump, at twice the pressure for which each pump is designed. For example, the comparable discharge for the pump in Figure 3-17 would be a total of 264 gpm (1000 L/min) at 290 psi (2000 kPa).

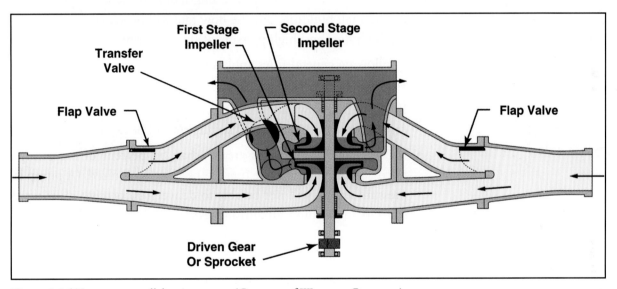

Figure 3-16 Two-stage parallel series pump. (Courtesy of Waterous Company)

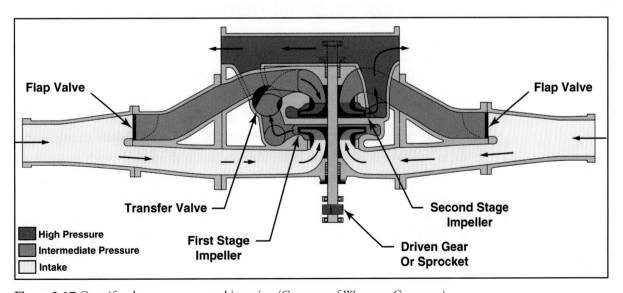

Figure 3-17 Centrifugal pumps connected in series. (Courtesy of Waterous Company)

The advantage of this type of pump is its ability to operate either at maximum volume or at maximum pressure within a close range of speed. High pressure is obtained in series at the same or slightly higher speed than is needed for capacity delivery in parallel. There are a number of ways to develop a higher pressure with the centrifugal pump. Single-stage pumps can be designed for high-pressure service by increasing the impeller diameter or the rated speed. This creates an impeller that is narrower and taller than normal. Alternatively, two or more impellers and casings may be assembled on one shaft as a single unit, forming a multistage pump; the discharge from the first stage enters the suction of the second stage, and so on. Many fire apparatus pumps are multistage (usually two-stage), thus permitting the most effective operation at the lowest possible engine speed. These pumps can deliver water against a higher pressure but at a lower flow; there is no free lunch, so to speak. Firefighters may need this higher pressure to overcome long hose length or to overcome higher elevations such as a tall building, but the flow rate will be lower. They cannot have both high flow and high pressure at the same time.

The following general rules apply for the use of the volume-pressure transfer valve on a two-stage pump:

- If the desired flow rate is less than 50 percent of the pump's rated capacity, the pressure (series) setting should be used.

- If the desired flow rate is greater than 50 percent of the pump's rated capacity, the volume (parallel) setting is required. (A pump in pressure setting will not be able to deliver 100 percent of its rated capacity under most circumstances.)

- The transfer valve should not be moved at any time when water is moving rapidly in the pump. The engine should be at idle while the valve is shifted. This will prevent internal water hammer.

Pump Locations and Drives

Fire pumps may be mounted at the front or rear of the apparatus or midship (about midway between the front and rear wheels). On most fire department pumper apparatus the pump is midship because that location is convenient and power is readily accessible. Since the pump is located behind the driving transmission, its power can come from the apparatus's main drive shaft. One method of doing this uses a split-shaft arrangement, in which a driving gear is set into the main drive shaft with a clutch or power-train that will disengage the rear of the drive shaft and engage the pump gears (Figure 3-18). In another method, the main drive shaft runs directly through the pump, and the impellers are engaged by a clutch or power-train. A third method of powering a midship-mounted fire pump is through a power take-off drive (Figure 3-19). In this arrangement, the gears within the transmission case drive the pump drive shaft. This permits the pump to operate either while the apparatus is stationary or moving.

Mounting the pump at the rear of the truck provides tremendous flexibility in design, operation, and safety by moving the hose connections away from the operator. The rear-mount arrangement shown in Figure 3-20 uses a version of the split-shaft design to power the pump. Front mounting requires the pump to be driven either through a gearbox with a clutch on the front end of the motor crankshaft or by direct drive (Figure 3-21). Front-mounted pumps of 793 gpm (3000 L/min) capacity are the most common size used in this location, although larger sizes are available.

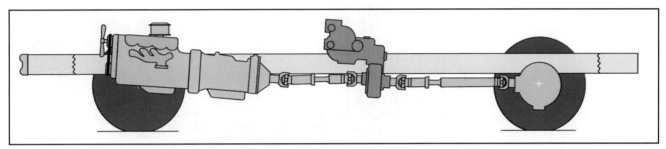

Figure 3-18 Fire pump mounted midship. (Courtesy of Waterous Company)

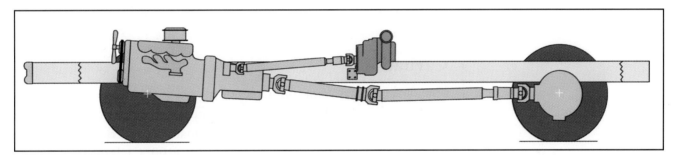

Figure 3-19 Pump driven through the PTO. (Courtesy of Waterous Company)

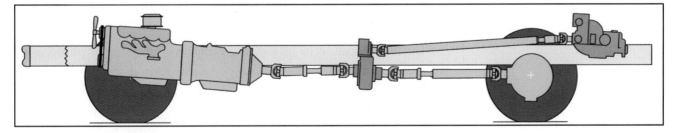

Figure 3-20 Rear-mount pump using a split-shaft design. (Courtesy of Waterous Company)

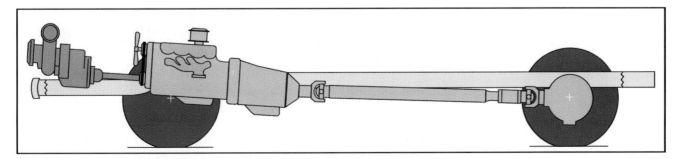

Figure 3-21 Front-mount pump. (Courtesy of Waterous Company)

An auxiliary engine also can drive the fire pump (Figure 3-22). Its drive-train arrangement can be the same as those for a front-mount pump or for a rear-mount pump through a transmission at the midship. The auxiliary engine can be skid-mounted, trailer-mounted, or built into the apparatus.

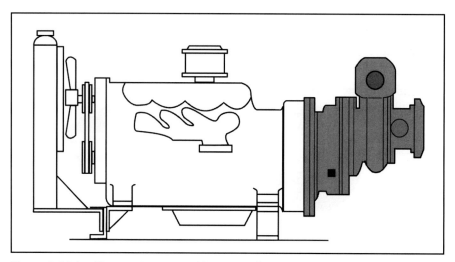

Figure 3-22 Auxiliary engine pump. (Courtesy of Waterous Company)

Cavitation

One of the fundamental principles of physics is that if the pressure on a liquid is reduced, its boiling point (the temperature at which it turns into a vapor) is also reduced. For example, at normal atmospheric pressure at sea level 14.65 psi (101 kPa) water boils at 212° F (100° C). If the atmospheric pressure were reduced to 10 psi (70 kPa), the boiling point of water would be 193° F (89.5° C). This is the underlying principle in the phenomenon called cavitation in fire pumps.

Cavitation can occur in any pump that is attempting to pump more water than is available. This can happen either from draft or at a hydrant. In the case of a centrifugal pump, when the pump "runs away from the water," bubbles of vapor form in the water near the impeller eye, where the vacuum in the pump is the greatest. As these vapor bubbles reach the discharge side of the pump, the pressure on the water becomes greater and the vapor bubbles implode. (The vapor turns back into water and other water rushes in to fill the rest of the space that had been taken up by the vapor). These implosions damage the impeller and housing by chipping away small bits of metal, eventually making the metal look pock-marked. The damage from cavitation can cause the impeller to become imbalanced, which subjects bushings, bearings, and shafts to strain and vibration.

When pumping at draft, cavitation may be caused by a partially clogged intake strainer, an intake hose that is too long or too small, a lift that is too high, water that is too warm, or a combination of these factors. When pumping from a hydrant, the cause of cavitation usually is either trying to pump more water than the hydrant can supply or using an intake hose that is too small.

The best cure for cavitation is to prevent it from happening. The vacuum gauges used on fire apparatus are not accurate enough to indicate whether cavitation is either imminent or happening. One of the best clues that a pump is cavitating is when an increase in engine speed does not produce an increase in pump discharge pressure. At the same time, the pump might sound as if pebbles were running through it. When these signs indicate that the pump is cavitating, reduce the engine speed. If more pressure is needed, partially close one or more discharge valves.

The Underwriters' label states the range of the pump when water is entering the pump at 0 pressure. From this information, you can determine the capabilities and limitations of the pumper in a wide range of pumping operations. For instance, at 150 psi, a 2,000 gpm pump will produce its rated capacity of 2,000 gpm (100 percent), but if it were necessary to increase the pressure to 200 psi to overcome line friction losses, then the flow would fall to 1,400 gpm (Table 3-1). As pressure increases, pump output drops; this is the nature of a nonpositive displacement pump. Using Table 3-2 you can see that a pump with a rated capacity of 4,000 L/min (100 percent) would produce 2,800 L/min at 1,350 kPa.

When a pumper is connected to a hydrant, the hydrant pressure does the work of supplying the pump; thus, more of the pump's power is available to produce flow and pressure. Also, the hydrant pressure will be added to the pressure developed by the pump. With the hydrant's help, the pumper can deliver its rated capacity at pressures of about 1,000 kPa. The limiting factors are the hydrant's flow and pressure capabilities and the pump's performance capability. If the residual pressure at the pumper is 200 kPa

Table 3-1 Flow reduction of a pump as pressure increases (Imperial).			
Rated Capacity	100% @ 150 PSI	70% @ 200 PSI	50% @250 PSI
1000 GPM	1000 GPM	700 GPM	500 GPM
1250 GPM	1250 GPM	875 GPM	625 GPM
1500 GPM	1500 GPM	1050 GPM	750 GPM
1750 GPM	1750 GPM	1225 GPM	875 GPM
2000 GPM	2000 GPM	1400 GPM	1000 GPM
2500 GPM	2500 GPM	1750 GPM	1250 GPM

Table 3-2 Flow reduction of a pump as pressure increases (metric).

Rated Capacity	100% @ 1000 kPa	70% @ 1350 kPa	50% @ 1700 kPa
3000 L/min	3000 L/min	2100 L/min	1500 L/min
4000 L/min	4000 L/min	2800 L/min	2000 L/min
5000 L/min	5000 L/min	3500 L/min	2500 L/min
6000 L/min	6000 L/min	4200 L/min	3000 L/min
7000 L/min	7000 L/min	4900 L/min	3500 L/min
8000 L/min	8000 L/min	5600 L/min	4000 L/min

Source: Alberta Fire Training School.

at a given flow, then the pumper can deliver that same flow at a pressure 200 kPa higher than it is actually rated for at that flow.

Examples: When a 2,000 gpm pumper hooked to a hydrant is flowing 2,000 gpm with a residual pressure reading of 29 psi at the pump intake, it will be possible to pump 2,000 gpm at 179 psi while the engine rpm is very close to that listed on the UL label (Figure 3-23a). Likewise, when a 4,000 L/min pumper hooked to a hydrant is flowing 4,000 L/min with a residual pressure reading of 200 kPa at the pump intake, it will be possible to pump 4,000 L/min at 1,200 kPa while the engine rpm is very close to that listed on the ULC label (Figure 3-23b).

Basic Fire Pump Maintenance

Only a fire pump that is properly operated and maintained can be depended upon to work in an emergency situation. As the bulk of the fire pump's use is in emergencies, a maintenance, inspection, and testing program that includes accurate written records is crucial.

The overhaul of a fire pump is costly and time-consuming. Regular maintenance, combined with common-sense practices, can do a great deal to extend the life of the pump.

Maintenance Schedules
After Each Use:

- Open and flush pump drains.

- Clean suction inlet strainers.

- Check oil levels.

- Reset governor or relief valves.

a.

Underwriters Laboratories of Canada Listed Fire Department Pumper ULC Label No. 45028		
Manufacturer's Registration No.		
Model No. **C800**	Date	**October 22, 1986**
Pump Rating		
4000 L/min	1000 kPa	2800 R.P.M.
2800 L/min	1350 kPa	3220 R.P.M.
2000 L/min	1700 kPa	3420 R.P.M.
Engine Manufacturer's No-Load Governed Speed: 3600 R.P.M.		
(P) Parallel (S) Series Operation, if applicable		

b.

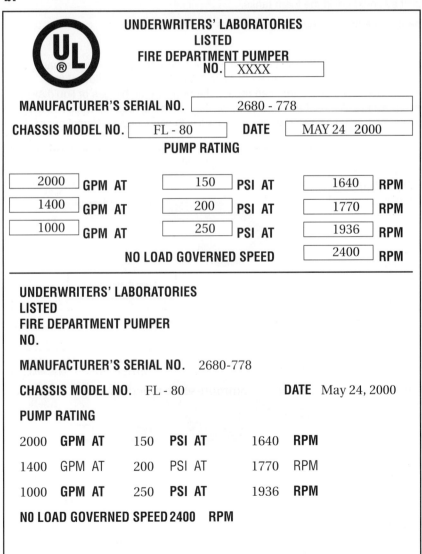

UNDERWRITERS' LABORATORIES
LISTED
FIRE DEPARTMENT PUMPER
NO. [XXXX]

MANUFACTURER'S SERIAL NO. [2680 - 778]

CHASSIS MODEL NO. [FL - 80]   DATE [MAY 24 2000]
PUMP RATING

[2000] GPM AT   [150] PSI AT   [1640] RPM
[1400] GPM AT   [200] PSI AT   [1770] RPM
[1000] GPM AT   [250] PSI AT   [1936] RPM
NO LOAD GOVERNED SPEED   [2400] RPM

UNDERWRITERS' LABORATORIES
LISTED
FIRE DEPARTMENT PUMPER
NO.

MANUFACTURER'S SERIAL NO.   2680-778

CHASSIS MODEL NO.   FL - 80   DATE   May 24, 2000

PUMP RATING

2000 **GPM AT**   150 **PSI AT**   1640 **RPM**

1400 GPM AT   200 PSI AT   1770 RPM

1000 **GPM AT**   250 **PSI AT**   1936 **RPM**

NO LOAD GOVERNED SPEED 2400   RPM

Figure 3-23 ULC pump certification plates. (Courtesy of fire etc. [formerly Alberta Fire Training School])

Foam Systems

TYPES OF FOAM

Fire departments use foams to extinguish fires. There are two types of foams, those used on Class A fires and those used on Class B fires. From Chapter 1 you will recall that Class A fires involve wood, paper, plastics, and rubber. Class B fires involve flammable liquids, such as gasoline, diesel fuel, and oils. The purpose of any fire-fighting foam is to keep oxygen away from the fuel and to cool the fuel. The Class A foam does this by draining water from its bubble structure in order to wet natural fuels; this is why Class A foam is often called a surfactant. By changing the surface tension of the water, this foam concentrate makes the water a more efficient extinguishing agent. Class B foams form a blanket over the fuel to keep the oxygen away from the fuel.

Foam systems generally extinguish fires from three to five times more efficiently than plain water. Therefore, less water is used to extinguish the fire, and cleanup is easier as less water damage occurs. Mixing the foam concentrate with the water at the correct rate is very important. This rate usually is expressed as a percentage, commonly ranging from 0.1 to 1.0 percent for Class A foam concentrate and from 1 to 6 percent for Class B. Generally, Class A foam costs less and is easier to clean up than Class B foam.

FOAM DELIVERY SYSTEMS

There are many different types of foam delivery systems. General maintenance guidelines for all foam delivery systems include:

1. Always flush the system completely after use.

2. Never mix different foam agents; they could gel in the foam tanks.

3. Never allow air to contact the foam concentrate in the storage tank, as a hard skin can form.

Eductor-Type Foam Systems

Eductor-type foam systems work on the Bernoulli principle, which states that any time a flowing gas or liquid is made to flow through an orifice, the speed of the gas or liquid must be increased (**Figure 3-24**). Speeding the flow of the gas or liquid in turn reduces its pressure. A properly designed system can reduce the pressure of this gas or liquid at the eductor to a level lower than atmospheric pressure. Atmospheric pressure will then push the foam up a tube to the eductor, where it will mix with the water and create foam. This same principle is used with sandblasters, Varsol guns, carburetors, and airplane wings.

Example: At the inlet to the eductor, the water pressure may be 200 psi (1378 kPa), and at the outlet of the eductor, the water pressure may be only 130 psi (896 kPa). At the center of the eductor, the pressure is low enough for atmospheric pressure to push foam concentrate into the eductor, where it is mixed with the water.

The bypass eductor is similar to the in-line eductor in that they both suffer from a pressure drop of 70 psi, from 200 psi to 130 psi (**Figure 3-25**). This pressure drop represents the energy that the eductor must use to operate properly. Most eductors will not function with an inlet pressure higher than 200 psi.

The around-the-pump–type of proportioning system cannot function with an inlet pressure higher than 10 psi (70 kPa) (**Figure 3-26**). That can be a problem if your pump is hooked to a water hydrant; you will have to gate down the incoming water pressure. This design requires the foam to go through the pump, and the foam bubbles could cause pump cavitation.

In-line, bypass, and around-the-pump eductors supply foam to every water pump outlet on the apparatus. To avoid this limitation, some departments install a smaller, PTO-driven water pump that is dedicated to foam operations. **Figure 3-27** diagrams a self-educting nozzle design often used for a water monitor installed on the top deck of the apparatus. Plain water remains available

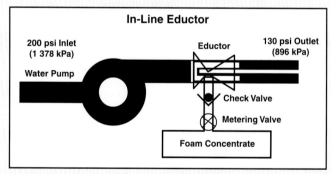

Figure 3-24 In-line eductor.

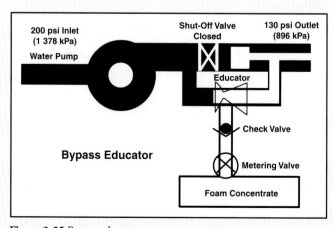

Figure 3-25 Bypass eductor.

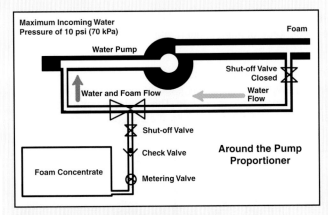

Figure 3-26 Around-the-pump proportioner.

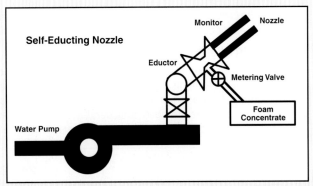

Figure 3-27 Self-educting nozzle.

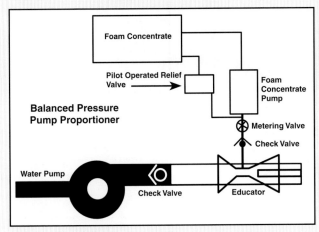

Figure 3-28 Balanced-pressure pump proportioner.

Figure 3-29 Foam concentrate injection pump. These pumps draw 20 to 60 amps of electricity.

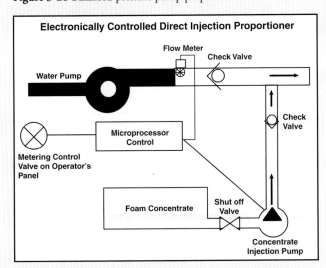

Figure 3-30 Electronically controlled direct injection proportioner.

Figure 3-31 Foam concentrate injection pump. Note the plumbing with valves.

to other appliances downstream from this pump. The balanced pressure pump proportioner system uses a positive displacement foam concentrate injection pump, usually of the piston-type design (**Figure 3-28**). The manually operated metering valve controls the amount of foam concentrate. A check valve ensures that water does not flow from the eductor back into the foam system. Any excessive foam concentrate is returned to the concentrate tank through a pilot-operated relief valve. This system is balanced because the pressure in the foam system is only slightly higher than the pressure in the fire pump. **Figure 3-29** shows an example of a common style of foam injection pump. The foam system shown in **Figures 3-30 and 3-31** also uses a dedicated pump to transfer the foam to the water but has the advantage of a microprocessor control for more accurate adjustment of the percentage of foam concentrate.

Compressed Air Foam Systems

A compressed air foam system (CAFS) uses a screw-type air compressor with a foam pump and water pump to produce a highly effective fire-fighting agent **(Figure 3-32)**. The use of compressed air has some excellent advantages. A very low percentage (in the range of 0.2 percent to 0.3 percent) of foam concentrate is needed. This lighter water-foam concentrate makes the fire hose considerably lighter and easier for the firefighter to handle. It also reduces line pressure loss caused by friction, allowing longer hose lays.

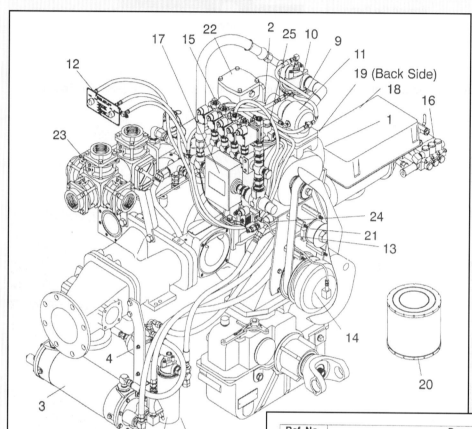

Figure 3-32 Compressed air foam system. (Courtesy of Waterous Company)

Ref. No.	Description
1	Rotary Screw Compressor (air end)
2	Encapsulating Sump Tank
3	Cooler Oil / Water
4	Oil Thermostat
5	Oil Filter
6	Water Strainer
7	Oil Drain
8	Water Drain
9	Air / Oil Separator Cartridge
10	Siphon Tube / Return
11	Modulating Inlet Valve
12	Auto Sync Control Panel (see Figure 2)
13	PolyChain® Belt
14	Pneumatic Clutch
15	Air Distribution Manifold with Electric Solenoids
16	FoamPro™ Proportioner
17	Air Flowmeter
18	Electric Relay Panel
19	Oil Fill and Level Sight Glass
20	Air Inlet Filter
21	PolyChain Tension Adjustment – Locking Bolt and Adjusting Bolt
22	Foam Manifold Built-in Check Valve
23	(500 GPM only) CAFS In-line Check Valves
24	Oil Temperature Sensor
25	Piloted Balance Valve

Compressed air, however, causes some maintenance concerns. First, the air gets hot when it is compressed. To cool the air and to lubricate the compressor, an oil mist is injected into the compressor and mixes with the air (**Figure 3-33**). Then this oil mist has to be removed from the air before the air can mix with the water and foam. This is accomplished by a sump tank, where the oil is collected (separated) and then sent to a cooler/thermostat/filter unit. The oil is cooled to remove the heat of compression and friction, then filtered and sent back to the oil injection port to repeat the cycle. If the oil overheats (above 250° F [121° C]), the air compressor drive clutch will disengage, and the water pump and foam proportioning system will continue to function normally. A sight gauge is provided on the oil reservoir/sump; it should indicate halfway up the window. Allow ten minutes for the oil to settle before checking the level. Check the oil level with the truck on level ground. The unit holds approximately 2.5 gallons (9.5 L) of an ISO grade 68 hydraulic oil. Ensure that the oil has an antiwear, antifoam, and antirust additive package. Change the compressor oil and filter after the first 30 hours of operation and then every year thereafter. Do not overfill with oil.

Typical compressor output is 80–90 standard cubic feet per minute (SCFM) (2265–2548 L/min) at 125 psi (861 kPa) with a maximum of 150 psi (1034 kPa). Standard conditions are defined as 14.7 psia at 60° F. To power a compressor of this size would take approximately 35 horsepower. This much horsepower could be converted to 89,000 BTUs of heat per hour, enough to heat a three-bedroom house when the outside temperature is −26° F (−32° C). The oil cooler removes this heat from the oil that cooled the compressor. The water-to-oil-tube type gets its water from the fire pump (**Figure 3-34**). If connected to a water hydrant or from draft, overheating of the water supply will not be a problem. If the truck is using its own water supply (booster tank), then the operator will have to very closely monitor the temperature gauges for overheating. It is possible to operate the air compressor only. Circulating water would still be needed to remove heat from the air compressor, and operational time would be limited to the amount of heat the water tank could absorb.

Figure 3-33 Darley compressed air foam system. Note the cutaway of the air filter.

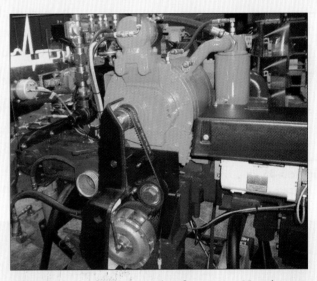

Figure 3-34 Waterous compressed air foam system. Note the green air compressor and the attached drive belt mounted above the red fire pump.

- Close drain valves.

- Flush systems that have pumped either salt water or corrosive water.

- Flush foam systems.

- Refill the booster tank.

- Clear air from the pump.

The regular maintenance of the fire pump should also include a weekly test in which water is discharged through the system.

Semiannually:

- Test the tightness of packing and gaskets by operating the priming system.

- Check vacuum leakage through valves.

- Check operation of the changeover or transfer valve.

- Test the pressure control device.

- Check the condition of the anodes. Fire pumps should be kept either completely full of water or empty when waiting for a call. Most departments keep the fire pump full. Some corrosion occurs even when the water's pH level is neutral, and as the water becomes more basic or acidic the rate of corrosion will increase. To combat this corrosion, anodes are installed. These anodes are made of a sacrificial metal and over time will dissolve in the water rather than the more expensive and more difficult to replace pump parts. The anodes may be made like those shown in Figure 3-35, or they may be part of the screens that cover the intake ports to the pump like those shown in Figure 3-36. How often they will need to be changed will depend upon the condition of the local water.

- Change the oil in the pump drive gearbox.

- Check the level and condition of the oil in the autolube reservoir if the pump has one. The autolube system is as simple as it is ingenious (Figure 3-37). It works on the hydraulic principle that you cannot compress a liquid. The diaphragm is exposed to intake water pressure, which could vary greatly, from a near vacuum to a very high

Figure 3-35 Plug-type pump anodes. (Courtesy of Hale Products, Inc.)

Figure 3-36 Intake screen anode on a Hale pump.

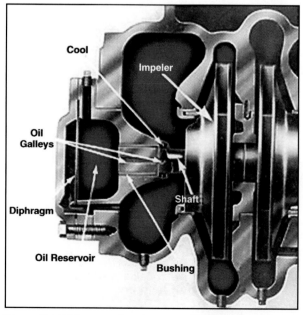

Figure 3-37 Diagram showing an autolube system for pump shaft lubrication. (Courtesy of Hale Products, Inc.)

pressure in the case of a relay pumping operation or a hot hydrant. The lubricant is trapped in the oil reservoir chamber. As the water pressure changes in the inlet of the pump, the diaphragm will allow a corresponding change in the oil pressure. This reduces the dynamic load on the shaft seal. As the shaft turns, centrifugal force throws out oil at the bushing around the shaft. This will cause oil to travel from the reservoir end of the shaft, down the shaft toward the shaft seal, and then back to the reservoir by passing between the bushing and the shaft, thus lubricating the shaft and bushing. This will work only if the reservoir is full and has no air pockets. Check the oil reservoir regularly for lubricant level and for signs of water contamination. Use the type of oil and grade recommended by the manufacturer.

Fire pumps should be tested annually to ensure the pump, driver, suction, and power supply all operate efficiently and to make any adjustments that may be required to return the unit to peak efficiency. (Fire-pump testing procedures are detailed in Chapter 7.)

The cooling and lubrication of a centrifugal fire pump are so dependent upon water that the pump must never run unless the pump casing is full of water. Pay close attention to the bearings and stuffing boxes during the first few minutes of operation to ensure they do not heat up or require adjustment. Read the suction inlet and discharge outlet pressure gauges occasionally to make sure that the inlets are not obstructed.

Make any required repairs and keep accurate records of all findings.

Routine Practices

- Use clean water only.

- Flush hydrants before filling the water tanks.

- Keep the suction strainer away from the bottom or sides of the static water source when drafting.

- Keep pump bearings and bushings properly lubricated.

- Do not run the pump without water.

- Change lubricants regularly.

Summary

Basic knowledge of the fire pump's operating principles will help you not only to perform routine maintenance but also to recognize unusual situations that might require special attention or repair.

Chapter 3 Review Questions

Short Answer

Write your answers to the following questions in the spaces provided.

1. The forward velocity force of flowing water through a discharge opening is referred to as _____ _____ .

2. List five properties of water that make it effective as a fire-fighting tool.

   _____

   _____

   _____

   _____

   _____

3. What are the six different types of water flow pressure?

   _____   _____

   _____   _____

   _____   _____

4. List three ways to create pressure for a water supply.

   _____

5. What is the most common type of fire pump in use today?

   _____

6. What are the two main parts of a centrifugal pump?

   _____

7. Explain the basic operation of positive displacement and centrifugal pumps.

   Positive displacement pump: _____

   _____

   _____

   _____

   Centrifugal pump: _____

   _____

   _____

   _____

8. Name two different types of positive-displacement pumps.

   _____

   _____

9. List three common uses of the positive-displacement pump.

_____

10. Why is the centrifugal pump considered to be a non-positive-displacement pump rather than a positive-displacement pump?

_____

_____

_____

11. Is a centrifugal pump self-priming? Explain.

_____

_____

12. What pushes water into the intake pipe of a pump?

_____

13. As a pump wears out, its ability to produce a given _____ against a given _____ will go down.

14. Which will a two-stage pump in series produce, high pressure or high flow?

_____

15. Which will a two-stage pump in parallel produce, high pressure or high flow?

_____

16. Name two situations when a department would want a two-stage pump.

_____

_____

17. What are the common names for the valve that can make a two-stage pump operate in either parallel or series mode?

_____

18. Name two common activation methods for the type of valve described in question 17.

_____

19. Intake valves must not open or close faster than _____ seconds. They must not open or close slower than _____ seconds.

20. Intake relief valves limit maximum inlet water pressure to the fire pump. This valve reduces the risk of what phenomenon?

_____

21. To where do the intake relief valve and the pump relief valve exhaust water?

_____

22. Why must outlet valves be designed to lock in any position from closed to fully open?

_____

23. Name three common locations for the fire pump to be mounted on the truck frame.

_____

_____

24. What causes cavitation?

_____

_____

25. List at least four jobs that should be completed after each use of the fire pump.

_____

_____

_____

_____

26. Why must a centrifugal pump never be run unless it is full of water?

_____

_____

_____

27. In addition to maintenance on the pump itself, what other components should receive regular maintenance and inspection?

_____

_____

28. How often should the pump undergo intensive testing?

_____

29. Name two reasons for the use of foam in preference to plain water as a fire-fighting medium.

_____

_____

30. Name the usual percentage of foam concentrate used for Class A foam.

_____

31. List three general guidelines for the maintenance of foam systems.

_____

_____

_____

32. Eductor-type foam systems have few moving parts, but they suffer from an inherent design fault. What is it?

_____

_____

_____

33. The around-the-pump proportion-type foam system generally will not operate if an intake pressure of 10 psi (70 kPa) or higher is present. True or false? If false, explain why.

_____

_____

34. The around-the-pump proportioning-type system allows the foam concentrate to flow around the pump. True or false? If false, explain why.

_____

_____

35. A balanced-type foam system uses a positive displacement pump to inject foam concentrate into the water. True or false? If false, explain why.

_____

_____

36. Name two advantages compressed air foam systems have over other foam systems.

_____

_____

Chapter 4
Basic Fire Apparatus Maintenance

Basic Fire Apparatus Maintenance

Introduction

All fire-fighting equipment, including fire apparatus, requires routine preventive maintenance to ensure it is kept in peak operating condition, ready for response at all times. This involves regular tests, inspections, servicing, and record keeping. Although the operator of the apparatus is generally accountable for the fire apparatus, apparatus and equipment maintenance/repair is the responsibility of the emergency vehicle technician. Only certified mechanics or technicians should undertake repairs to the apparatus.

Take the time to review manufacturer's specifications and become aware of the recommended guidelines for scheduling maintenance. Together with NFPA standards, they will help you to establish and maintain a regular schedule for daily, weekly, monthly, semiannual, and annual inspections. Keep accurate records of all of the results.

Basic fire apparatus maintenance includes everything from cleanliness and touch-ups to scheduling oil changes and keeping careful records of all repairs and inspections. This is the duty of the drivers/operators. Their role in the maintenance program does *not* include actual oil and filter changes or repairs to the vehicle. Preventive maintenance will pay off in the long run, both in the reliability of the apparatus and in the cost of repairs.

Cleanliness

The old adage that cleanliness is next to godliness might be overstating things a bit, but not much. In the fire service, public image means a great deal, and the appearance of the fire apparatus contributes much to that image. Children dream of riding on a shiny red (or yellow) fire truck, not on a dirty vehicle that happens to have fire-fighting capability. Clean the apparatus whenever dirt can be seen on it. Not only does a clean apparatus leave a good impression on the tax-paying public, but it also ensures efficient operation.

A clean engine and clean functional body parts allow for proper inspection. This same cleanliness keeps oil, moisture, and dirt from accumulating on the engine and maintains mechanical efficiency of the engine, its wiring, fuel injectors, and functional controls.

General Maintenance

To keep your apparatus in good operating condition, you must follow a regular maintenance program and adhere to your department's established maintenance schedule. You may find that developing a checklist proves very helpful. The information that follows is a general outline of the basic requirements common to most departments and most vehicles.

Daily

Conducting the following checks each and every day on the fire apparatus is important to ensure it will be ready to roll when the alarm sounds. If you detect any problem areas, the apparatus should be removed from service until the component has been restored to serviceability.

Driving Area

Check for broken, torn, or otherwise damaged seatbelts. Far too many firefighters die every year going to and from the fire while riding in the truck. In too many cases the seatbelts were inoperable or were not used. While you as the technician cannot force the firefighters to use the seatbelts, you can make sure the belts are operational and easy to get at (not stuffed into a seat or behind a bunch of junk).

- ☑ Check all windows and mirrors for cracks, cleanliness, and ease of operation.

- ☑ Ensure windshield wipers are intact and operable. Great advances in wiper design have made them work better at road speed. (A cheap set may work well at the curb, but that's not where you need them.)

- ☑ Make sure air, oil pressure, voltage, fuel, and water temperature gauges work.

- ☑ Pressure-test all brakes by operating the foot pedal.

  - *For gasoline engines with hydraulic brakes:* With the engine running, apply the foot brake. The pedal should feel firm and not move toward the floor. Now, shut off the engine, again apply the foot brake, and restart the engine. If the unit has a power brake assist (most will), the foot brake should now move toward the floor, stop, and again stay firm and not fall toward the floor.

- *With an air brake type system:* Build the air pressure up to at least 100 psi (689 kPa), turn off the engine, and apply the foot brake. If you cannot hold the brake pedal all the way down by yourself, use a wooden stick or other commercially available device to hold down the brake pedal. Now, while the brake pedal is depressed, walk around the vehicle; you are listening for any leaking air at both the front and back brakes. The air brake system is allowed by law to have some leaks. The FMVSS (121) standard (in Canada, the CMVSS standard) is that after a one-minute stabilization period you are allowed to lose 2 psi (13.8 kPa) during the next two minutes on both air tanks. With a tractor-trailer unit with the trailer connected, you are allowed 4 psi (27.6 kPa) per tank. The photograph in Figure 4-1 shows a corroded air brake pot that would not pass the above test.

Figure 4-1 Air brake pot with corrosion in the spring-brake section. **CAUTION:** The spring inside this section is highly compressed and extremely dangerous if not properly caged. (Courtesy of Phil Wagner)

Example: Consider a single-axle truck whose tanks have 120 psi (828 kPa) of air. With the engine shut off, apply the air brakes with your foot. After one minute (the standard stabilization period) the gauges read 105 psi (724 kPa). During the next two minutes you should not lose more than 2 psi (13.8 kPa) from each tank. By reading the gauge on the dash, it is very difficult to see the difference between 115 psi (793.5 kPa) and 112 psi (772.8 kPa), which would be more than 2 psi (13.8 kPa) and, therefore a failure, but it is very easy to hear a 2 psi (13.8 kPa) leak when you walk around the truck and listen.

While losing 2 psi (13.8 kPa) in 2 minutes may be legal, that is not the level of maintenance you want for your emergency fire apparatus. You should endeavor to stop all leaks. While the brakes are applied, walk around the vehicle and check the angle between the slack adjuster and the push rod. This angle must not be less than 90° when the brakes are fully applied.

If the angle is less than 90°, say 70° (Figure 4-2), then much of the force the air chamber has produced will not be delivered to the slack adjuster and therefore will not be available to stop the truck. The requirements of NFPA 1901, 10-3, for the performance of the service brakes are very clear. They must be capable of bringing a fully laden apparatus to a complete stop from an initial speed of 20 mph (32 km/h), in a distance not exceeding 35 feet (10.7 m) by actual measurement on a substantially hard surface that is free of loose material, oil, or grease. This standard is not easy to meet. A fully loaded vehicle must have full water tanks and the equivalent weight of each firefighter who would normally travel in the vehicle. That weight is typically calculated as 250 pounds (114 kg) for each firefighter. This allows 200 pounds (91 kg) for the firefighter and 50 pounds (22.7 kg) more for the equipment, SCBA, and other tools. In short, a truck with an empty water tank and only a driver might pass this test, while the same truck fully laden might not.

Figure 4-2 A brake badly out of adjustment. Clearly the angle formed between the push rod and the brake adjuster is less than the 90° minimum.

CAUTION:

It is not recommended practice to conduct the service brake test with people on board.

While the new generation antilock brakes (ABS) are a great improvement over standard brakes and even over the first-generation antilock brakes of the early 1970s, they are no good if not maintained. Most, if not all, ABS systems today come equipped with self-adjusting slack adjusters. As with any mechanical device, however, there is a possibility of failure. If the adjusters fail to adjust for the normal wear in the brake drums and shoes, then the ABS systems can actually work against themselves and defeat their purpose. During a normal brake application the ABS computer receives wheel-speed input from the wheel-speed sensors at each of the wheels. If one wheel starts to slow down faster than the rest (to lock up, as occurs on a slippery road surface), the computer will signal a valve at the wheel that is slowing down to exhaust air pressure from that brake pot. This will ensure that all wheels connected to the ABS will slow down at nearly the same rate, which in turn will ensure that the driver can maintain steering and therefore control of the vehicle. But the system will not compensate for automatic slack adjusters that are defective and out of adjustment.

Exterior

Check the tires for proper inflation, alignment, cuts, and wear and tear. Marks on the sidewall of the tire will indicate the maximum psi allowed and the maximum weight allowed on each tire when the tire is cold (this is defined as room temperature). The maximum weight per tire is critical on fire apparatus because, unlike a gravel truck that would be unlikely to spend even 50 percent of its service life fully loaded, a fire truck spends almost 100 percent of its time fully loaded. If the tire is run at a pressure as little as 25 percent less than its maximum, its service life will be greatly reduced.

Example: If a tire's recommended maximum inflation were 100 psi (690 kPa), you could run the tire with as little as 85 psi (586.5 kPa), and it would still have a reasonable service life. But if you were to let the pressure fall as low as 70 psi (483 kPa), the tire's life would be reduced.

If you allowed a tire's pressure to fall too low, the steel belts in the tire sidewall would flex to such an extent that were you to reinflate the tire, the sidewall would fail (Figure 4-3). This is called zippering and can result in serious injury. When a tire is found to have been run underinflated, it must be removed from the vehicle, dismounted, and professionally inspected (Figure 4-4). Remember that you cannot check tire pressure with a hammer or with your foot: use an accurate tire gauge (Figure 4-5).

Figure 4-3 A zippered tire caused by being driven with low air pressure. When the sidewall flexed on the underinflated tire, the steel cords broke. Then when the tire was inflated, the sidewall burst. Note the almost new tread. The standard inflation pressure for this tire when cold is 105 psi (724.5 kPa) if used as a single wheel or 95 psi (655.5 kPa) if used as a dual. This tire was used as an inside dual, and the tire pressure was only 50 psi (345 kPa) when the technicians tried to inflate it to the correct pressure using a clamp-on tire chuck.

Note: While there is no NFPA standard on the use of recapped tires, many departments use them for rear tires but not for steering tires. If your department decides to use recapped tires and you want the same tire casing back, consider branding your tire casings. There is an area on the side of the tire where you may apply your brand. Check with the recapping company to find out more about this method of tire identification.

Figure 4-4 Tire cage, an essential piece of safety equipment for tire maintenance.

Figure 4-5 The correct use of a tire gauge.

Figure 4-6 The above wheel has many problems. First, this is a BUD type wheel (also called a ball-seat mount) and the nut at the 12 o'clock position is the correct fastener for this type of wheel. The other fasteners are called hub pilot mount and are not meant for this rim. Second, note the black spaces around the BUD type nuts; these indicate that this rim has been loose at one time and is now non-serviceable. Also note the rust on the bottom right-hand corner of the rim; this would indicate that the wheel has been weakened.

☑ Inspect wheel studs, valve stems, and lug nuts or wedges to ensure all is as it should be (Figure 4-6).

☑ Check the floor for fluid leaks: coolant/antifreeze (green or blue), power steering fluid (red), brake fluid (clear), and engine oil (dark colored).

☑ Ensure all visible and audible warning signals are operational. These would include low-air-pressure warning lights and buzzer and may also include a low-pressure wigwag.

☑ Check lights for cracked lenses, loose attaching screws, and exposed wires.

☑ Check for a tight fit on all doors. Check hinges and latches for complete closure.

☑ Make sure all water tanks are full. This does not mean simply to see if the tank lights show an appropriate level but to actually get on top of the tank and look in.

☑ Operate the changeover valves on two-stage pumps to make sure they function properly.

Engine

☑ Check the oil level (Figure 4-7). *If the oil level is too high,* the oil will hit the crankshaft and cause an imbalance problem that will most likely break the crankshaft and cause severe engine damage. *If the*

Figure 4-7 Example of an oil level indicator mounted on the side of the engine oil pan. This type of indicator is becoming more popular on fire apparatus.

oil level is too low, this could indicate oil consumption or a leak: fix this problem right away. The old method of looking at the oil and changing it "because it looks black" is not reliable.

The oil level on the dipstick should be halfway between the add and full marks. On a large diesel engine of 300–500 Bhp, the difference between these marks is normally one U.S. gallon (4 liters). If your oil level is slightly above the add mark and your department buys its oil in 1-gallon jugs, should you pour in the whole jug and be a quart/liter over the mark or do nothing and wait until the oil level drops to the add mark? Being a little over is better than a little under: add the whole jug. Remember this is an engine that holds 10 gallons of oil, so being a quart over is not so bad. The best solution is to convince your department to stock the right brand of oil in 1-quart/liter jugs. While using oil in 1-quart/liter containers for top-up might seem slow and costly, it may actually save you time and money. Often you can buy oil in quart/liter containers on sale at a price comparable to bulk. Using oil from sealed quart/liter containers also ensures that clean oil goes directly into the engine and does not get contaminated by a dirty funnel or container.

Make sure you document any oil you add to an engine. In a large department many personnel could be involved with servicing the apparatus. Consult with the engine manufacturer on oil change intervals, and use the manufacturer's recommendations as a starting base for the change period. Then follow a program of oil analysis to determine the correct change period for your brand of oil, filters, and driving conditions.

☑ Verify that the radiator has an adequate level of coolant (Figure 4-8). Make sure the engine is cold before you do this; if in doubt, touch the upper radiator hose. If the engine is at running temperature and you remove the radiator cap, the coolant will instantly boil and serious injury will result. Many newer trucks come with an expansion tank, often on the side of the radiator. These work well if the system is maintained. However, if the coolant system has even a small leak, the expansion tank cannot function properly. The purpose of the expansion tank is to allow a small amount of the total coolant to flow into it as the coolant expands when heated. When the system cools (as it does after the engine has been shut down for a time) this coolant will be drawn back into the radiator. This system prevents corrosive oxygen from entering the engine block and causing damage to the engine. When the engine is at room temperature, the radiator must be full of coolant and the expansion tank filled only to its *cold* mark.

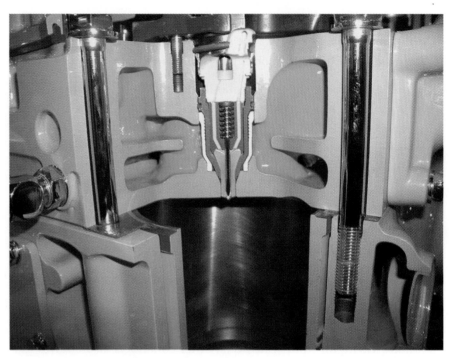

Figure 4-8 A cutaway of a Caterpillar diesel engine showing how little room there is for antifreeze coolant (green) around the cylinder walls of the engine. Coolant is the most often forgotten fluid in an engine.

CAUTION:

Never remove a radiator cap on a hot engine.

Note: Use only active-spring-loaded hose clamps as shown in Figure 4-9 (see page 101). They cost more than the other types, but they do not leak as much.

☑ Check all batteries and connections as well as the master switch.

☑ Check the fuel level. Most departmental rules stipulate the tank should be filled if the fuel level is at or below the three-fourths mark. Again, find out your department's policy. NFPA 1901, 10-3.4, spells out the requirements for fuel tanks. The tanks must carry enough fuel for at least 2.5 hours of pumping at 150 psi (1,035 kPa) net pump pressure and at the suction conditions specified in this standard or at 60 percent of gross engine horsepower for 2.5 hours, whichever is greater. While installing two fuel tanks may seem prudent, this generally is not recommended. This is because the driver(s) may find it difficult to find the tank changeover switch while driving or during pumping operations. Also the fuel for gasoline-powered engines can go stale if left in a tank that is not regularly used.

Weekly

Once a week, the apparatus must undergo additional, regular maintenance. (Now you see why the checklist comes in handy.)

☑ Check the levels of the:

- *Transmission oil.* Your apparatus will most likely have an Allison automatic transmission. This transmission has served the fire service well, but it does require regular maintenance, and the best place to start is the oil level. A common mistake that most inexperienced people make is to check the transmission oil when the engine is cold. The engine must be running at idle, the transmission in neutral, and the park brakes set. The transmission oil must be between 160° F and 200° F (71° C and 93° C) to be able to give a correct reading. If you were to check the oil after the apparatus had been shutdown over night and the engine had not yet been started, the oil level would be way above the operating level. While this is not the correct time to check the oil, it does indicate that there is oil and that it is the correct color.

> **Note:** Few things can destroy an automatic transmission, but one is engine antifreeze (glycol) in the transmission oil. An automatic with antifreeze in the oil will fail; it is just a matter of time. This can happen very suddenly, as with an overnight oil cooler failure. So the need to check both the engine-coolant level and the transmission-oil level and color cannot be emphasized enough. If antifreeze has leaked into the transmission oil, the transmission will have to be dismantled and inspected and, at the very least, the clutches replaced. A sharp operator can save the department a very expensive repair, but most importantly your department may have time to bring up a reserve unit before the damaged unit fails on the way to a fire.

Most departments use Dexron III oil, which is also called Mercon. This oil replaced the older Dexron II automatic transmission fluid (ATF). It should be bright red and not smell burnt. If in doubt, compare its color and smell to new oil fresh from a jug. Not all departments will use Dexron. For more severe use, such as fire trucks that operate off road or have severe hills in their operational area, your department may explore the specifications of a C-4 transmission oil. Many top quality diesel motor oils meet Allison specifications for a C-4 oil; however, not all motor oils are rated C-4, so be careful. One major advantage of using C-4 transmission oil is that the same oil used in the engine can also be used in the transmission. This reduces your stock of oils and may allow you to buy in bulk with a financial savings to the department. The major advantages of C-4 over ATF are its ability to hold up at a higher temperature and the fact that it comes in many grades of viscosity, one of which may match the grade of motor oil you use now.

There are many different types of oil top-dressing (these are also called oil additives.) In most cases they contain friction modifiers that help to reduce friction, most often a type of Teflon or an extreme pressure additive. While the reduction of friction no doubt would be a great thing for mankind, the transmission requires a certain coefficient of friction (C of F) to provide the clamping force to properly engage the clutch packs. Any reduction in the C of F will allow the clutch packs to slip under load or during engagement; this will shorten the life of the automatic transmission.

- *Differential oil.* In recent years there has been a move to use synthetic gear oils; in most cases this oil is used as the fill oil at the factory. If your differential requires the API GL-5, then you should use synthetic oil matching the grade of oil you need (i.e., 80W–140). These oils provide greater oil-film strength at high temperatures, a wider temperature range, and most importantly, superior resistance to high-temperature oxidation and thermal degradation than standard mineral-based gear oils.

- *Power steering fluid.* While ATF is often specified for power steering systems, some instances require power steering fluid or even motor oil.

- *Brake fluid.* When checking fluid levels on vehicles with hydraulic brakes, do not leave the cover off the brake fluid containers or master cylinder. The brake fluid is hygroscopic, which means it wants to absorb moisture from the air. Do not leave the top off a master cylinder or brake fluid container any longer than is essential. If you find a container of brake fluid whose cap has been off more than ten minutes, do not use it. If you find a master cylinder that has a loose cover, you will need to drain and flush the system. Be careful: brake fluid and paint do not mix; brake fluid is highly corrosive to paint.

☑ Check the brake system (when equipped with hydraulic brakes). Many manufacturers recommend changing and flushing the brake fluid at least once every two years. The many different types of brake fluid have changed in the last few years; make sure you use the correct type. Check for leaks of brake fluid from wheel cylinders and hoses.

☑ Check air brake systems for leaks. Bleed moisture from the air tanks. This task is often done weekly, but that may conflict with many state and provincial laws that require the driver to drain air tanks daily. While many trucks are equipped with air driers, remember that they have a life cycle of about six months in normal use. If you counted on the air drier to remove the moisture and the drier were to fail, your air tanks could begin to fill with water. This would have two effects. The first would be to reduce the amount of compressed air your tanks could hold and therefore reduce the amount of stopping power you have. The second would be that the moisture would freeze if the truck were used on a cool day (0° C or 32° F), and the truck might not be able to move.

☑ Check battery terminals and cables to ensure adequate contact.

> ## CAUTION:
> **Be careful! There are at least three different types of brake fluid. Make sure you use the correct one for your system.**

☑ Check belts (fan, generator, and alternator) to confirm tightness and physical condition. This step is even more important on emergency apparatus as these vehicles have larger than normal electrical demands. Many highway trucks with far more engine horsepower may be able to satisfy their electrical loads with an alternator of only 125 amps, but a fire truck will normally have an alternator of 250 amps or more. This puts a tremendous strain on the drive belts.

☑ Operate valves in the auxiliary cooling system.

☑ Check drains and hose connections for security, working condition, and foreign objects within (Figure 4-9). One of the most important performance tests for a pump is the dry-vacuum test. If a pumper cannot pass this test it will not be able to pump from draft. One of

Figure 4-9 Air-to-air exchanger for the engine turbocharger. Spring loaded hose clamps are used here; they are more effective than screw-type clamps (commonly used in automobiles) as they apply a constant holding pressure. Note also the blue, high-temperature silicone hose that is used.

the more common reasons that pumpers fail this test is the inability to close off the drain lines. This in turn prevents the primer pump from creating the needed vacuum inside the pump body. Some departments do not put a high priority on this. The reasons they give are that they only pump from a hydrant or the on-board water tanks, but they have a false sense of security. Even in the most developed of urban areas, you may have to draft from a swimming pool, small pond, or flooded ditch.

☑ Check the drive shaft and universal joints.

☑ Start the motor and observe oil pressure when the engine is idling at operating temperature. An engine will have a high oil pressure when cold, but this pressure will fall when the engine is warmed up. The pressure must not fall below 20 psi (138 kPa) when the engine is warm.

☑ Clean underneath the chassis. To determine how best to do this you will have to consult OSHA, EPA, and other federal, provincial, and state pollution regulations. It is very important to keep the underside of the apparatus clean, but at the same time you need to avoid a conflict with pollution laws. The days of high pressure or steam washing of undercarriages may be over in some jurisdictions, but the job still needs to be done. Doing it may come down to a bucket and a soapy rag. Minor leaks must be found and corrected before they become major leaks and the cause of vehicle component failures.

☑ Clean the engine and electrical motors.

☑ Check for loose pins, nuts, and studs (Figure 4-10). Always use new cotter pins; never reuse a cotter pin. While you usually will reuse bolts, always use new lock washers. Make sure that any replacement bolts are of the same grade and thread as the originals. Beware of counterfeit bolts. These bolts may display the proper head marking indicating the bolts' grade, but they do not meet the standards. If the price seems too good to be true, it probably is.

☑ Check fuel sediment bowl (if so equipped). The sediment bowl should be clean with no sign of rust or water (Figure 4-11).

Periodic

Although a qualified technician or mechanic will make all required repairs, the firefighter or driver/operator should be aware of both when they are required and when they are done so as to maintain the records and to ensure the work is completed on schedule.

☑ Change air filters as required. The construction and over-the-road trucking industries use an air inlet restriction device that is mounted in the cab and visible to the driver. These devices warn that a high

Figure 4-10 This picture shows the displaced pin that the truck cab hinges on when raised. The retaining bolt for this pin has fallen out and allowed the hinge-pin to migrate to the right. (Courtesy of Tom Gaines)

Figure 4-11 A glass sediment bowl on a fuel system full of rust and water. (Courtesy of Phil Wagner)

vacuum exists between the air filter and the engine. As the filter becomes plugged by dirt, the vacuum will increase. As the air filter plugs, engine performance will decrease, fuel consumption will increase, and exhaust smoke may also increase and become very dark. These indicators work only if there is an air filter. NFPA 1901, 10-2.4.1, states that the air inlet shall be protected so as to prevent water or a burning ember from entering the air intake system. If a burning ember were to enter an air intake during a grass fire for example, the paper filter element would be quickly destroyed. The filter would never plug, and therefore this gauge would never perform as intended. This is another special aspect of the fire service. While the over-the-road operators may rely on the air restriction device to warn of a plugged air filter, emergency vehicle technicians must check air filters manually and not rely solely on the air inlet gauge.

☑ Drain and refill the crankcase and change oil filters at least twice a year. Your oil change interval should be based on engine hours or fuel consumed. If you have not reached the recommended number of engine hours or amount of fuel consumed, you may want to change oil twice a year. This may happen most often in rural departments. Modern motor oils contain antirust additives that absorb and neutralize acids in the oil. These acids are formed from running the engine for short periods. Any time an engine is to be stored for a long time, the manufacturer will always require that the oil and filter are changed and the engine is run up to operational temperature before it is shutdown. This means that the new oil has had time to coat the engine parts with an antirust additive ensuring that the engine will not seize up during storage.

☑ Check the fuel pump periodically. Look for both flow and pressure specifications on a gas engine.

☑ Perform tune-ups on gasoline and diesel engines annually. After any major engine repair, NFPA 1901 recommends performing a service test on the pump; this would include an engine tune-up.

☑ Lubricate the chassis, distributor, starting motor, water circulating pump, generator, and steering gear, as required.

☑ Service-test the apparatus, as required.

Equipment

Fire apparatus maintenance also includes the equipment that is carried on the vehicle but not permanently attached to it. Most of this equipment is fairly standard. To keep it in good working order, you should follow a scheduled maintenance program.

Daily (Or after Use)

☑ Check all portable extinguishers for damage or loss of pressure.

☑ Check hose loads to make sure they are ready for quick deployment when needed.

☑ Inventory all nozzles and appliances.

☑ Operate all valves.

☑ Inspect protective breathing apparatus for air pressure and proper operation.

☑ Ensure all hand lights and flashlights are in working order.

Weekly

☑ Start all portable engines or motor-driven equipment and allow them to warm up to make sure they are operating efficiently.

☑ Clean all appliances.

Periodic

☑ Take all hose from the apparatus to check its condition and then reload, changing the bends. Service-test hoses annually.

☑ Hand-clean all portable ladders.

☑ Lubricate all motor-, gear-, or hand-operated appliances as outlined by the manufacturer's manuals.

☑ Flush the water pump weekly and the water tank monthly.

Electrical Components

Once the cleaning and exterior maintenance are complete, you must turn your attention to the electrical components of the apparatus. Moisture or corrosion may damage or render inoperative many electrical connections on a fire apparatus. As a result, it is important that you conduct a thorough inspection of the electrical components.

Lighting

Visibly check to verify that all components of the lighting system are working. Have all burned out lamps or flashers replaced immediately. Some of the components you should inspect include:

- Headlights
- Dimmer switches
- Clearance, parking, brake, and back-up lights
- Compartment lights and switches
- Warning lights and switches
- Floodlights and switches

Electrical Motors

Regularly make sure that each of the electrical motors on the apparatus is operational. The best way to test them is to turn them on and observe what happens. Be sure to include:

- Rotating lights
- Hose-reel rewind
- Windshield wipers
- Apparatus controls
- Heater/defrost fans

Battery

The battery is a unique part of the apparatus that requires special attention (Figure 4-12). The battery stores electricity for future application. However, it is more than just a storage receptacle for power. The battery actually generates electrical energy from the chemical reaction produced when two contrasting materials, such as the positive and negative plates, are immersed in an electrolyte.

Figure 4-12 Three of the six batteries used on a truck with a Cummins ISM engine. This engine has a displacement of 661 cubic inches. Note that the battery box is of stainless steel construction. These batteries have a reserve capacity of 180 minutes each; therefore, the total reserve capacity of the six batteries is 1080 minutes, or 18 hours.

Basically, batteries are composed of five elements:

1. A plastic or hard rubber case encompasses the unit.

2. Positive and negative plates made of lead are housed within.

3. Plate separators made of porous synthetic material prevent short circuits. A positive and negative plate, combined with a separator, form one element or cell.

4. Electrolyte, also referred to as battery acid, is a solution of water and sulfuric acid.

5. The positive and negative lead terminals serve as the connection points between the battery and whatever it electrifies.

The battery operates in a constant process of charge and discharge. When you start a vehicle, you draw power from the battery, leaving it weaker. Conversely, the alternator of the vehicle recharges the battery when you drive without accessories. Over time, the battery can become weak and incapable of delivering electricity at a useful voltage.

☑ To help maintain a full charge on your batteries, routine maintenance is very important.

☑ Inspect the battery case for cracks and loose or leaking terminals.

☑ Visually inspect the battery terminals, clamps, and the terminal ends of the cables for corrosion. Clean, if required. The recommended method for cleaning battery terminals is to use a commercial terminal cleaner (or a mixture of baking soda and water) and a wire brush.

When the terminals are cleaned they may be painted with special protective paint (black for negative and red for positive). Vaseline or grease may also be used to coat the outside of the connections and protect them from the corrosive effects of air.

CAUTION:

The acid in the corrosion from battery terminals will burn exposed skin.

☑ Ensure the battery box and hold-down devices are secure, clean, and in good repair.

☑ Where a battery compartment is provided, make sure it is adequately ventilated to prevent buildup of heat and explosive fumes. If the battery is located in the engine compartment, ensure the heat shields are in good condition.

☑ Check the cell electrolyte level (water levels). Add *distilled* water, if necessary.

☑ Check specific gravity.

☑ Test for voltage.

☑ Recharge, when required.

In the event your battery becomes discharged, you can recharge it by feeding electrical current in a reverse process. A full charge will restore the battery's ability to deliver its full power. Follow these six steps whenever you charge a battery.

1. Identify the positive and negative grounds of the battery.

2. Attach the red cable to the positive terminal.

3. Attach the black cable to the negative post.

4. Connect the charger to a reliable power source.

5. Set the desired battery charging voltage and charging rate, where equipped.

6. Reverse the procedure to disconnect the charging cables.

CAUTION:
Batteries produce explosive hydrogen gas when being charged.

Many inexpensive chargers do not provide a way to reduce the voltage as the battery reaches a full state of charge. It is possible with one of these chargers to overcharge the battery; this would damage the battery (if it were not defective before, it will be now). Chargers that do reduce the voltage are called constant potential chargers (*potential* being another word for voltage). These chargers are expensive but worth the money; if you do not have one of these chargers, use a digital voltmeter when charging a battery. Normally you can charge a 12-volt battery at any amperage as long as you never exceed 15 volts. If you do exceed 15 volts, the battery will begin to gas and have an acid smell.

A battery's becoming discharged to the point that it will not start the engine is referred to as a deep cycle. Every time this happens, it shortens the battery's life. You must find out why this has happened and correct the problem. Batteries for fire apparatus are specified in NFPA 1910, 11-4 as *high-cycle* type batteries. Even though they can withstand more deep cycles than a normal automotive battery, their service life is still shortened.

Almost all new equipment will come from the factory with a sealed maintenance-free battery of the group 31 type. This battery does not need to have water added, nor should you try to do so. It performs well and has a high-cycle rating. Batteries not only have to be able to start the engine but must also provide a source of electrical power. This is reserve capacity and is measured as the battery's ability to deliver 25 amps for a specified number of minutes without falling below 10.2 volts. That is the standard battery reserve capacity test for the automotive industry, however, and does not apply to fire apparatus, which may have a minimum continuous load well above 100 amps.

Example: A battery is rated at 120-minutes reserve capacity. That means it will deliver 25 amps of current for 2 hours and not fall below 10.2 volts. But what if a fire truck needs 100 amps to continue to run? Because the truck needs four times as many amps as a car, the length of time this capacity is delivered is reduced by a factor of four, giving approximately thirty minutes reserve capacity.

The other battery rating that is important to fire apparatus is cold cranking amps (CCA). This is a measure of a battery's ability to deliver vast amounts of current to start the engine. It is determined by how many amps the battery can deliver for 15 seconds without falling below 9.6 volts. How large does this rating have to be? A simple guideline is three CCAs for each cubic inch of engine displacement for a diesel and two CCAs for a gasoline engine.

Example: Let's take an 8V92 DDA engine. This engine has 92 cubic inches per cylinder and eight cylinders, so it has a total of 736 cubic inches. Multiply 736 by three and you have determined that the engine needs a battery capability of 2208 CCAs. Three group 31 batteries at a rating of 750 CCAs each will give you the needed cranking energy.

The more important of these two methods of measuring a battery's performance is reserve capacity. This is for the simple reason that fire apparatus are normally stored indoors or at least in a location where the temperature does not go below 32° F (0° C). Temperature has a dramatic effect on both a battery's performance and its service life. As a battery's temperature goes down, its chemical action slows, and therefore its current output also falls. A battery should not be stored partially charged in a wet, warm place, but of course these are exactly the conditions found in a fire apparatus. Not only is the battery not allowed to become fully charged because of the constant stop-and-start driving with long periods of idle, but it is also exposed to high temperatures either under the hood or in the battery boxes. As a battery's temperature rises, its ability to produce current goes up, but its rate of self-discharge and plate deterioration also increase. This explains why the same size batteries might have three to four years of useful life in a grader or backhoe and have trouble lasting two years in a fire truck.

Onboard Battery Conditioners

Parasitic battery drains that cannot be eliminated may necessitate installing a battery conditioner. Parasitic drains are caused by electrical loads

Note: A 6-volt battery does not produce more power than a 12-volt battery. If either two 6-volt batteries or one twelve-volt battery would fit in a battery box with the same dimensions (width, height, and length), the one 12-volt battery should outperform the two 6-volt batteries for two reasons. First, each of the 6-volt cases will have one extra sidewall, and these two sidewalls will take space where battery cells could be located. Second, two extra external battery cables will be needed to hook up the 6-volt batteries; the additional cables and connections will lead to more voltage drops and more places for corrosion.

that cannot, or should not, be turned off, such as engine and transmission computers that need to retain their memories and rechargeable radios and flashlights that are in their holders so they will be fully charged when needed. These drains can be excessive and, over time, kill the batteries that are used to start the engine.

Battery conditioners (often referred to as float conditioners) are far more than just inexpensive trickle chargers that can be bought at the tire store. Not only will they maintain the battery at a full charge, but they will also ensure that it does not overcharge. Remember that battery conditioners will not compensate for a defective battery, nor are they intended to be used as battery chargers. While these conditioners are of a high quality, they can very easily give the technician a false sense of security. Onboard battery conditioners will not save you money if you do not have in place a proper program of battery maintenance and replacement. Follow the manufacturer's guidelines for installation and remember that because it uses 120 volts it falls under the rules of line voltage power systems and must be installed following the regulations of the National Electrical Code.

Voltage Regulator

The regulator requires special maintenance. However, you should be familiar with indications that signal the need for correction or repair. It is important to remember that voltage regulators regulate only voltage, not amperage. The voltage regulator's job is to maintain the voltage at a set maximum.

Example: Let's say that the set maximum voltage is 14.4 volts for a 12-volt system. You would want no less than 13.8 volts measured at the battery, and a well-maintained system would provide closer to 14 volts. Where did the other 0.4 of a volt go? It was lost between the alternator and the battery. This is called voltage loss, and it happens in even the best-maintained systems. The aim is to reduce it as much as possible.

Any time current flows in a wire, the wire offers some resistance to that flow. If the current is too high or the wire is too small for the flow, the wire will get very hot. As a wire heats, its resistance goes up and its ability to conduct current goes down. This becomes a vicious circle. To ensure that this does not happen, NFPA 1901, 11-1, addresses the following requirements:

- The wire must be 125 percent larger than the size normally needed to power the device.

- Voltage drops will not exceed 10 percent of the applied voltage.

- Because most of the electrical problems you will encounter will be at the connections, the use of star washers is strictly forbidden.

- When connecting wires, it is very important that you make the connection correctly.

- Use the right sized wire stripper to remove the insulation.

- Never use a knife to strip wire, no matter how careful you think you are. If the knife cuts or nicks even a few of the wire strands, the wire's conducting abilities are reduced.

- Use a crimp connector that does *not* have a plastic cover; if a crimp connector does have a plastic cover, remove it.

- Use the proper crimper and then solder the connection. Most people fail to understand that the crimp provides only the mechanical connection while the solder provides the electrical connection.

- Cover the connection in one of several ways: with electrical tape, with shrink tubes that seal around the connection and remove all the air when they are heated, or with a liquid electrical sealant that hardens after it is applied.

Indications

- The battery indicates a full charge and the ammeter displays low charging rate under normal circuit load.

- The ammeter shows low charge rates when the battery is at or near full charge. This is normal, as an ammeter does not indicate how much amperage the alternator is producing; it indicates only how much is going to or from the battery.

Example: Think of a battery as a bank account. You must put the money in before you can take it out. A battery is similar: batteries do not produce current, they only store chemicals, and the chemical reaction can produce this current (electron) flow. So the next time your battery fails to start your engine, don't ask why no current came from the battery; ask what went wrong with the way you put in the current. In other words, the ammeter tells the operator only that the battery is running at a surplus or a deficit.

Because of the uniqueness of fire apparatus electrical systems, any repairs to the electrical system should be carried out only by personnel specifically trained on these systems.

Alternator

Which term is correct: *generator* or *alternator*? In the early years (1920s) the device that produced the electron flow was called a generator. Generators had some faults in that all the amperage they could produce had to flow through their brushes. Even though in most cases it was barely 40 amps, this amperage shortened a generators' life. Generators were high-maintenance units. During the early 1970s the prices of the first solid-state components came down so that using them in new devices called alternators became practical. Alternators had only four moving parts: two brushes and two bearings. The brushes had to carry only the field current, which in most cases was 4–8 amps; therefore, they had a much longer service life. Not until the late 1990s were alternators called generators again. They really are alternators, even though they are sometimes called generators.

Both alternators and generators require special maintenance. The fire service places very different demands on an alternator than does the over-the-road trucking industry. Alternators in excess of 250 amps are now common on fire apparatus. Why are they needed? The alternator must be sized to provide for the apparatus's electrical demand at an ambient air temperature of 200° F (93° C); this is a typical temperature within an engine compartment. The alternator's performance must be tested at both engine idle and fully governed engine speed. The need for these two tests reflects the nature of the work the apparatus must perform. An alternator will not be able to supply its full rated output at engine idle, but the truck may need to power its full electrical load while at idle. If this electricity cannot come from the alternator, then it must come from the batteries, and that would cause the system voltage to fall. Normal system voltage will be close to 14.4 volts for a 12-volt system when the engine is running. A reduced alternator output under the same electrical load demands will lower the system voltage and could cause radios, lights, and other important devices to fail or perform erratically. With the introduction of electronic engines and transmissions, the system voltage level has become even more important. The nonelectronic engines and transmissions could continue running with no electrical system, but the new electronic systems cannot. They will shut down if a voltage of less than 10.3 volts is detected. For this reason NFPA 1910, 11-14.3.3, requires that a system's voltage does not fall below 11.7 volts (23.4 volts for a 24-volt system) during tests for alternator performance at full load.

Alternator Test at Full-Governed Engine Speed A low-voltage warning device must be installed to warn the operator if battery voltage falls below 11.8 volts for a 12-volt system or 23.6 for a 24-volt system. This device must give an audible and a visual warning. Run the engine for two hours at full-governed speed with the total continuous electrical load applied. If the alarm sounds within this time, the test is considered a failure and repairs or modifications need to be made.

What would cause an alternator to fail this test? The short answer is heat. As the ambient temperature under the hood increases, an alternator's ability to produce amperage will fall. Again you must consider the nature of fire fighting. The fire truck does not have the advantages of an over-the-road highway truck. The highway truck gets tremendous cooling from the road draft. This road draft will move great amounts of heat from the engine compartment; however, fire apparatus must be designed to operate without this road draft. Some apparatus do not take this into account, and this two-hour test will reveal these design mistakes.

Alternator Test at Idle Now run the engine at its normal idle speed. With only the minimum electrical loads activated, monitor the battery voltage and ammeter. If the battery shows a discharging condition, then the test is a failure. Remember that an ammeter does not show the output of the alternator; it only indicates if the batteries are charging or discharging. Each department must determine its own minimum electrical load requirements according to its standard operating procedures.

Engine

An engine in good operating condition is vital to any fire-fighting operation. Not only does the engine get you to the fire scene, but it also supplies power for much of your equipment. If the engine does not work, neither does the fire pump, aerial ladder, or elevating platform.

One sure way to guarantee you contaminate the oil, waste fuel, and cause unnecessary engine wear is to start the engine simply to warm it up. Suffice it to say that manufacturers recommend that you do *not* start an engine for this reason. If you really do not believe that the engine will start, then you need to have a long talk with the people who perform your maintenance. Besides, running the engine one day does not guarantee it will start the next time the alarm sounds. If the battery is charged and the fuel tank full, a properly maintained engine should have no problem turning over when it is called into operation.

Your role in maintaining the integrity of the engine is to ensure that it receives regular service inspections. Like your car, the fire apparatus requires routine engine maintenance to keep engine failures to a minimum. One part of that service inspection includes lubrication. Proper lubrication of the engine does a great deal to avoid out-of-service time and large repair bills. It is a basic of good maintenance that helps to keep the engine running smoothly. Consult the manufacturer's guidelines and the mechanic to select the grade of lubricant best suited to the requirements of the apparatus.

A certified mechanic should do all repairs and adjustments to the engine.

Chassis/Driveline Components

A fire apparatus must be able to execute its intended service under what are often adverse conditions that may require operation off paved streets or roads. Always remain alert to changes, vibrations, noises, and other signs of potential problems with the chassis. Regularly inspect the chassis for defects, structural integrity, perforations, and missing or loose parts (Figure 4-13). At the same time, recognize that driveline components installed in the chassis are not immune to problems and require a degree of maintenance equal to that of the engine. To avoid service problems, it is important to regularly inspect the following:

Figure 4-13 A badly cracked frame. (Courtesy of Dan Cromp)

- Master cylinder brake fluid reservoir

- Compressor belts

- Power-steering hydraulic fluid reservoir

- Brake pedal for floorboard clearance

- Brake pedal of hydraulic systems while driving

- Tires for air pressure, wear, and breaks

- Universal joints. Unlike a standard over-the-road truck, a fire truck will have an extra set of U-joints because of the transfer case that powers the fire pump. All joints must run at a certain angle. Even if it were possible to design a perfectly straight driveline, it would not be advantageous as it would not have a long life. A joint must operate at an angle because this causes the needle bearings inside the joint to rotate against the cross section ever so slightly. This small amount of movement ensures that the needle bearings get a new charge of grease.

- Balance weights on the drive shaft (Figure 4-14). It is imperative that these weights are in place to ensure low vibration and long bearing life. In past years the nonelectronic Allison transmission would reach

Note Driveshaft Weight

Note: To grease a U-joint, use good quality grease. Wipe the grease nipple with a rag and flush the end of the grease gun with new grease. Now grease the U-joint until new grease comes out of *all four* cross seals, not just a few seals. Do not worry that this will blow out the seals; they are designed to be greased in this manner. If you cannot get grease to come out of all four cross seals, then disassemble the cross. You will probably find that the cross that has not been getting grease has been damaged and the whole U-joint needs to be replaced. Before conducting a stall test on an Allison transmission, ensure that you personally check all U-joints and steady bearings, as this stall test will stress these components.

Figure 4-14 Balance weights on the driveshaft.

a ratio of 1:1 in high gear. This meant that the driveline would rotate at the same speed as the engine. With the new electronic transmissions and their overdrive top gears, the driveline now rotates much faster. At these faster speeds, driveline angles, and balance become even more important than in past years.

- Steering devices, axles, and springs. Any signs of rust on these components indicate a weakening of the metal (Figures 4-15, 4-16, and 4-17). If rust is excessive, components may need to be replaced. Check with the manufacturer for replacement guidelines.

- Tail pipe and muffler

- Parking brake adjustment

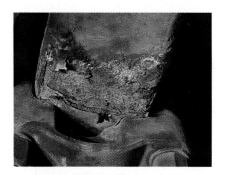

Figure 4-15 Rust on a drive axle. (Courtesy of Phil Wagner)

Figure 4-16 Rust on a steering axle. (Courtesy of Phil Wagner)

Figure 4-17 Rust on a steering axle spring. (Courtesy of Phil Wagner)

Fire Pump

Pump overhaul is expensive and time consuming. Excessive wear and damage that might necessitate overhaul can be avoided or delayed by following a routine maintenance program. In addition to regular testing, regular preventative maintenance should be conducted on fire pumps to detect potential problems, especially on aerial apparatus where the pump is not used as often.

The only way to ensure a fire pump works is to operate it. It is not necessary, however, to run the pump at full capacity. The following checks are a few generic guidelines for routine maintenance. Refer to the manufacturer's recommendations for specific instructions.

- Ensure the pumps are mounted securely. Use only the correct grade of bolt (beware of counterfeit bolts) and never reuse a lock washer.

- Open and clean all pump drains weekly or after each use.

- Inspect and clean all suction intake strainers weekly or after each use.

- Check the pump gearbox (also called the transfer case) and priming pump reservoir for proper oil levels and traces of water weekly or after each use. If you find water in the pump gearbox, it is not enough

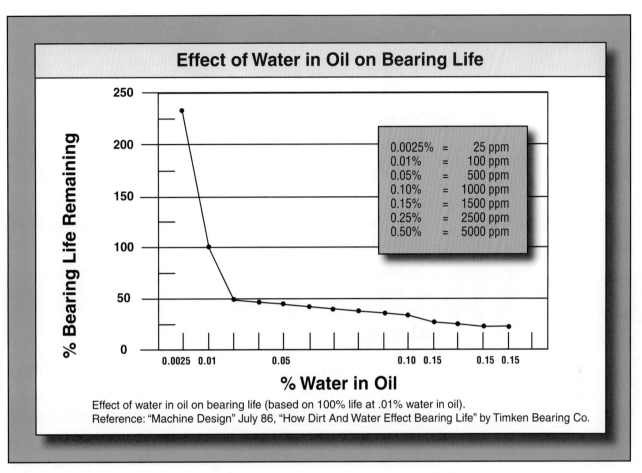

Figure 4-18 Effect of water in oil on bearing life. (Courtesy of Parker Filtration Manual)

just to drain out the oil and replace it with fresh oil. You must find out how the water got into the gearbox. One of the most common causes of this is an improperly adjusted water pump seal packing. This seal is right above the drive shaft that powers the gear box, so any water that is leaking out of the water pump seal will fall onto this drive shaft. If the amount is within specifications, the seals will remain intact and no water will get into the gearbox. If the amount is too large, the seals will be flooded and some water will get past the seals into the gearbox and mix with the lubricant. How much water is too much? Even a small amount of water shortens the life of a bearing. According to the Parker Filtration Manual, the oil will not have a milky appearance until at least 300 parts per million (ppm) of water are present, or 0.03% water by volume. At this volume the damage will have been done and the life of the bearing reduced by 60 percent (Figure 4-18). Water will reach the saturation level at about 1000 ppm, or 0.10 percent, at which time the bearing life will be 20 percent of new. The chart in Figure 4-18 further shows how the life of bearings is reduced when water gets in the oil.

- Reset the governor or relief valve weekly or after each use.

- Trigger the changeover valve while pumping from the booster tank weekly.

- Inspect packing glands for extreme leakage. Replace packing or gaskets as required.

- Operate all valves (including the relief valve) weekly.

- Check all gauges to ensure they work.

- Operate the pump primer with all pump valves closed semiannually.

- Remove discharge caps and check vacuum leakage through valves semiannually.

- Recalibrate the flow meter according to the manufacturer's instructions.

Fuel Tanks

Full fuel tanks are critical. NFPA 1901, 10-3.4.1, states that fire apparatus should contain a fuel tank with a minimum capacity to allow the apparatus to run for at least 2½ half hours when operating at rated pump capacity and pressure, or to operate at 60 percent of gross engine horsepower for 2½ hours, whichever is greater.

It is impossible to predict how long an engine will have to run during an emergency situation. During long operations, fuel tanks often have to be refueled time and again. Ensure that the apparatus always starts out with a full fuel tank.

Water Supply Tanks

Always keep water tanks full to help prevent corrosion within the tank. Flush the tank and pump regularly to reduce the likelihood of stagnation and rust within. The frequency of flushing depends on local conditions; as a minimum, flush the system with clean water at least twice a year, more often in areas where excessive deposits build up. When and if rust particles appear in the water, empty the tank, rinse it out, and refill it to full capacity with clean, fresh water.

Aerial Devices

Aerial devices must be systematically maintained with special attention to the operating parts. Dependability is crucial to lives—namely yours and those of your coworkers. Both aerial ladder and elevating platform devices require detailed and labor-intensive maintenance. Suffice it to say that regular routine maintenance, such as lubrication and oil changes, needs to be augmented by a program of maintenance specific to the aerial device and its components. For our purposes, consider the following recommendations for aerial device maintenance:

- Check the torque of bolts connecting the turntable to the frame. This will require a torque wrench capacity of 400 to 500 foot-pounds of torque (1.77 to 2.23 KN). Never rely on an air wrench for proper torque. If you find one broken turntable bolt, suspect all the bolts. As a bolt is torqued it gets longer; this stretching is what gives the bolt its tremendous clamping force. If the bolt has been overtorqued, it will become deformed, and although it may not break instantly it will break in the future. Use the correct grade of bolt. One trick that is often used to help detect any problems is to paint both the bolt head and the turntable. In the future, if you notice the bolt head has broken the paint, then it must be loose.

> **Note:** Which clamps stronger with the same amount of torque, a bolt with fine threads or a bolt with coarse threads? Let's consider a ½-inch SAE (Society of Automotive Engineers) grade 8 bolt. This bolt can be torqued to 92 foot pounds (0.4 KN). But which would clamp the metal tighter under that amount of torque, fine threads or coarse? The answer is in the number of threads per inch. The coarse-thread bolt would have 13 threads per inch and the fine-thread bolt would have 20 threads per inch. Thus, the coarse threads give a mechanical advantage of 13:1 while the fine threads give an advantage of 20:1. For the same given torque (92 foot pounds), the coarse threads would clamp with 1196 pounds of force (92×13), while the fine threads would clamp with 1840 pounds of force (92×20).

- Ensure all locking and retaining devices are complete and operational.

- Inspect all cables and sheaves; replace where necessary.

- Make sure all balls, races, and sliding surfaces are lubricated properly.

- Check the tubing and hose within the hydraulic system for damage.

- Inspect the boom structure for cracks and deformed parts.

- Ensure the gear teeth on the turntable are clean and properly lubricated.

- Look for bent flanges or worn sleeves in the sheaves.

- Check the tightness of all attachments, rails, hinges, and leveling sheaves.

For more information on maintaining these devices, refer to the instruction manual delivered with the aerial apparatus or to NFPA 1904, *Standard on Aerial Ladder and Elevating Platform Fire Apparatus.*

Record Keeping

An accurate record-keeping system that details all maintenance checks, tests, and repairs is crucial to good fire apparatus maintenance (Figure 4-19). The system should be simple and should work for the department. Otherwise, it is difficult to keep up, hard to use, and worse to read.

Keep a detailed inventory of the equipment carried on each apparatus. Create checklists to guide you through your regular maintenance chores (Figure 4-20). At the same time, record all routine tests as well as preventive maintenance records, including shop checkups, servicing, and repairs and parts replacement (Figures 4-21 and 4-22).

Summary

The necessity for a fire apparatus maintenance program is obvious. An alarm is not the time to discover that your apparatus has a flat tire or is otherwise inoperable due to negligence. Keeping apparatus in a constant state of readiness helps to ensure the reliability of equipment and reduces the cost and frequency of repairs. Additionally, accurate schedules and records of the fire apparatus maintenance can help to pinpoint problem areas as well as to speed annual tests.

Apparatus Data and History Card

Public Property No.

13-

Basic Data

Make		Type		Year
First Cost	Manufacturer's No.		Title No.	Specification No.

Length	Ft. In.	Width	Ft. In.	Make of Engine	GVW Rate	Cu. In. Displacement	Cylinder
Fuel Tank		Tire Size	Tire Type	Battery	Clutch	Transmission	Wheel Base	Radio 12V 6V

Pumpers

| Pump Make | | Pump Type | | Pump Volume | | Water Tank |

Ladders

Height		Type
Complement of Ladders		

Year	License Plate	Status	Condition of Body	Condition of Pump or Ladder	Motor Hours	Road Mileage	Annual Pump Test Result	Times Out of Service	Down Time	Cost of Repairs Non-Accident		Cost of Repairs Accident	
										Labor	Material	Labor	Material

Figure 4-19 Apparatus data and history card.

_____Fire Department

"Three Month & Annual"
Maintenance Checklist

Service to be performed:
- ❏ "A" Three months or 1,000 miles
- ❏ "B" Annual or 4,000 miles

Date_____
Vehicle & make _____
Apparatus no. _____
Odometer reading _____
Engine miles _____
Pump hrs. since last report _____
Total pump hours _____

		❏ Adjustment Made		❏ Repairs Required
A	**B**	**Items** ❏ **O.K.**		**A B**
		1. Check "Each Week" reports		7. Chassis
		2. Cooling system		• Lubricate chassis
		• Hoses		• Wheel bearing (inspect and repack)
		• Radiator core (leaks)		• Universal joints and flanges
		• Flush (add rust inhibitor)		• Clutch and linkage (automatic transmission)
		3. Oil		• Oil all linkages
		• Drain and refill crankcase		• Steering mechanism
		• Replace oil filter		• Springs (damage and alignment)
		• Road transmission oil level		• Springs — U-bolts
		• Pump transmission oil level		• Muffler and tailpipe
		• Rear Axle oil level		• Axle, flange nuts (tighten)
		4. Brakes		• Wheel lug nuts (tighten)
		• Handbrake		8. Body of Apparatus
		• Inspect and adjust brakes		• Cleanliness, paint
		• Hydrovac — inspect and lubricate		• Doors and windows (hinges, etc.)
		• Drain air tank		9. Lights
		• Master cylinder (fluid level)		10. Siren and bell
		5. Battery breakdown test		11. Windshield wipers (blade, arm, etc.)
		6. Engine		12. Pump mounting bolts (tighten)
		• Spark plugs (clean, etc.)		13. Winterzation
		• Distributor cap, rotor, points		14. Auxiliary generators, pumps, etc.
		• Starter & generator		15. Road test
		• Voltage regulator		• Brakes
		• Set ignition timing		• Steering
		• Adjust carburetor at idle		• Clutch
		• Choke and Throttle		• Transmission
		• Manifold nuts (tighten, etc)		• Engine performance
		(1)_____ (2)_____		• Engine oil pressure (hot)
		(3)_____ (4)_____		1) Idle 2) 1,000 rpm
		(5)_____ (6)_____		• Engine temperature
		(7)_____ (8)_____		• Rattles
		(9)_____ (10)_____		
		(11)_____ (12)_____		
		• Adjust valve lash		Mechanic's Signature
		• Clean air and fuel filters		
		• Instruments and gauges		Officer in Charge
		• Check fuel pumps		

Figure 4-20 Three-month and annual maintenance checklist.

Rockford Fire Department
Preventive Maintenance Work Sheet

"A" Service	"B" Service
To be performed each 1,000 miles or every 4 month period	To be performed each 4,000 miles or every 12 month period

Date _____ Vehicle No. _____ Company No. _____

"A" Service	"B" Service				"A" Service	"B" Service	
		1.	Check driver's report				26. Drain air tank
		2.	Fan belt adjustment				27. Springs for broken leaves and alignment
		3.	Cooling system hoses				28. Muffler and tailpipe
		4.	Radiator				29. Tighten spring U-bolts
		5.	Starter and generator Commutator and brushes				30. Tires — cuts, pressure, unusual wear
							31. Tighten wheel — lug nuts
		6.	Distributor cap — rotor and breaker points				32. Tighten axle flange nuts
							33. Master cylinder fluid level
		7.	Generator regulator				34. Fifth wheel
		8.	Tighten manifold nuts Check for leaks				35. Road and pump transmission level
							36. Rear axle oil level
		9.	Engine compression reading 1.____ 2.____ 3.____ 4.____ 5.____ 6.____ 7.____ 8.____ 9.____ 10.____ 11.____ 12.____				37. Drain and refill crankcase
							38. Replace oil filter if necessary
							39. Clean air and fuel filters
							40. Oil all linkages
		10.	Inspect and regap spark plugs				41. Wheel bearings — inspect and repack
		11.	Adjust valve lash				42. Inspect and adjust brakes
		12.	Set ignition timing				43. Lubricate chassis
		13.	Adjust carburetor at idle				44. Flush cooling system — add rust inhibitor
		14.	Battery breakdown test				45. Door hinges, latches, window glass regulators
		15.	Radiator shutters				
		16.	Hand brake				46. Check for oil and water leaks
		17.	Lights				47. Cleanliness of apparatus
		18.	Siren and bell				48. General appearance of paint and metal
		19.	Windshield wipers				49. Road test — check brakes, steering, operation of clutch, transmission, engine performance, rattles
		20.	Instruments and gauges				
		21.	Choke and throttle				
		22.	Pitman and steering arms drag link and tie rod				50. Engine oil pressure (hot) Idle — 1,000 rpm
		23.	Clutch pedal clearance and linkage				51. Engine temperature
		24.	Universal joints and flanges				52. Check auxiliary generators, pumps, etc.
		25.	Hydrovac — inspect and oil				

Figure 4-21 Preventive maintenance worksheet (page 1). (Courtesy of Rockford [Illinois] Fire Department)

Preventive Maintenance Work Sheet

Each item adjusted shall be explained under "remarks"
Example: Tail light bulb replaced — burned out.

REMARKS: _____

_____

_____

_____

_____

_____

_____

_____

_____

_____

_____

_____

_____

_____

_____

_____

_____

_____

_____

_____

_____

_____

_____

_____

_____

_____

Figure 4-21 Preventive maintenance worksheet (page 2). (Courtesy of Rockford [Illinois] Fire Department)

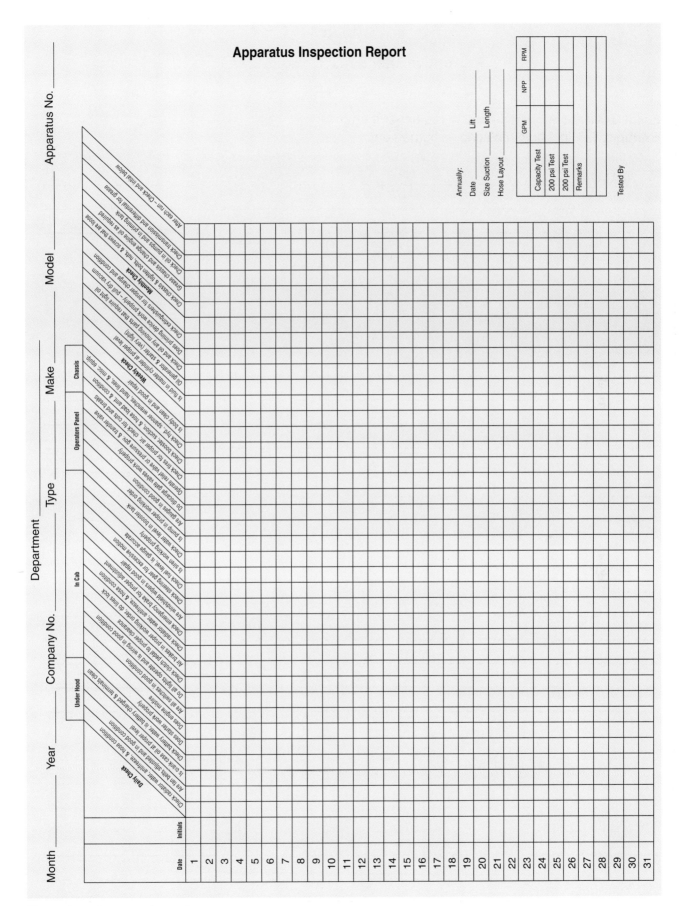

Figure 4-22 Apparatus inspection report.

Chapter 4 Review Questions

True or False

Write *True* or *False* before each of the following statements. Correct those statements that are false.

_____1. Fire pumps are not used as often on pumpers as on aerial apparatus.

_____

_____

_____2. It is a good idea to start the engine of fire apparatus every day to warm it up.

_____

_____

_____3. Apparatus operators are responsible for carrying out some apparatus maintenance.

_____

_____

_____4. Maintenance should be performed according to manufacturer's guidelines.

_____

_____

Short Answer

Write your answers to the following questions in the blanks provided.

1. Explain the difference between maintenance and repair.

_____

_____

2. List at least five tasks that should be performed daily as a part of routine apparatus maintenance.

_____

_____

_____

3. List at least five tasks that should be performed weekly as a part of routine apparatus maintenance.

_____

_____

_____

_____

4. List, in order, the six steps for attaching battery charger cables to a vehicle's battery.

   a. _____

   b. _____

   c. _____

d. _____

e. _____

f. _____

5. List a minimum of three electrical motors on the apparatus that should be checked periodically.

_____

_____

_____

Chapter 5
Lubrication and Cooling System Maintenance

Lubrication and Cooling System Maintenance

Introduction

Correct and adequate lubrication is the basis of an effective maintenance program for any fire department apparatus. Failed lubrication programs are the major cause of apparatus failure. Predictive maintenance depends upon an understanding of lubrication, or *tribology*, and the special requirements of fire department vehicles. A good predictive maintenance program uses oil analysis, heat analysis, and vibration analysis and an excellent tracking system (sometimes called bookkeeping) to determine if the lubricants and filters used are effective for the vehicle, when the oil and filters need to be changed, the amount of wear on lubricated components, and when components need to be changed before they fail.

Lubrication plays a vital role in the following functions:

- Friction reduction

- Heat removal

- Contaminant suspension

- Sealing

The choice of a lubricant that will perform all of these functions must take into account the particular use of the vehicle in which the lubricant will be used. This chapter provides an overview of the special lubrication requirements of fire service vehicles, an evaluation of motor oils and greases suitable for fire service vehicles, a description of the information obtained by fluid analysis, and case studies that illustrate the value of fluid analysis in a predictive maintenance program. It also includes a section describing how heat and vibration analysis can be used to monitor the effectiveness of the lubrication program and the wear on system components, and thus is used to predict and forestall any major component failures. A detailed glossary of lubrication terms is included on the companion CD-ROM.

Finally the problems created by an undetected electrical current running through the coolant are discussed and maintenance methods to prevent such problems are outlined.

Motor oils

The traditional motor oil grade for diesel engine use is 15W-40. This oil has served the fire service very well, but new, more expensive synthetic motor oils are now available. Are they worth the added expense, and will they protect as well in hot weather? The new 0W-40 motor oils for heavy duty (HD) diesel engines do, in fact, protect fire department engines better in hot weather than does 15W-40. To understand this higher performance level, let's examine the development of motor oils for HD diesel engines.

For many years the oil of choice for HD diesel engines was straight-weight motor oil, and that choice was usually 40-weight oil. This oil was very thick in cold weather, but when brought up to operating temperature in the engine it had the correct viscosity to protect the engine parts. It became possible during the late 1970s to formulate, at reasonable cost, a multigrade oil (15W-40) that would last in a four-cycle diesel engine. This oil would flow in cold weather like 15-weight oil but would thicken in high temperatures to act like 40-weight oil. How was this possible? It was achieved with the use of polymers.

The best way to explain polymers is to describe them as looking like very small octopuses. In cold temperatures these polymers shrink (think of the arms of the octopus closing up to the body and thus reducing the overall size of the octopus). The oil containing these polymers can flow very easily in cold weather. At high temperatures these polymers stretch out and are able to form longer chains of polymers. As these chains form, the oil viscosity increases and the oil flows as if it were thicker. Motor oils are composed of approximately 80 percent base-stock and approximately 20 percent additives; the polymers are in the additives.

Multigrade oils are not recommended for two-cycle diesel engines. This is because the two-cycle diesel engine operates differently than the four-cycle diesel engine. The cycles of the four-cycle engines are intake, compression, power, and exhaust. During the intake cycle there is very little if any pressure on the piston rings. As the piston moves down on this intake stroke, the engine piston rings can wipe oil from the cylinder walls and recharge their supply of lubricating oil. The piston of a two-cycle engine, on the other hand, has no intake cycle; it has only a power cycle and a compression cycle. Its top piston rings are always under load, and the resulting extreme pressure between the ring, piston, and cylinder walls shears the oil polymers. (The camshaft of a four-cycle engine rotates at only half of engine crankshaft speed, while the camshaft of a two-cycle engine turns at the same speed as the crankshaft. This faster rotation of the camshaft gears and camshaft followers in the two-cycle engine breaks down, or shears, conventional polymers.)

When the polymers are sheared they can no longer expand or contract in response to changes in oil temperature. Remember that 15W-40 oil is really a base stock of 15-weight oil that uses polymers to make it perform like 40-weight oil at high temperature. If the polymers have been sheared (destroyed) because of these higher temperatures and shearing contact between the camshaft and its followers, the oil will revert to its base stock of 15-weight. This 15-weight oil is far too thin to protect engine components, and increased engine wear results. Thus, only straight weight (40-weight, 50-weight) motor oils are recommended for two-cycle diesel engines.

As two-cycle engines are no longer available on new fire apparatus, your fleet will most likely contain both two-cycle engines and four-cycle engines. With recent developments in polymer design, some oil manufacturers are now producing motor oils suitable for these mixed engine fleets. Be careful, however, and obtain a written guarantee of performance from the oil manufacturer for these oils.

Even with the best base-stock crude oil and an excellent additive package, the widest spread attainable in a multigrade oil is about 20–25 points (for example, 10W-30, 0W-20, 15W-40. Getting a wider spread would necessitate completely breaking down the crude oil at the molecular level and reassembling it to perform as needed. This has now been achieved, but the process is expensive. However, it does make very pure oil with uniform properties. With the correct additive package, it is now possible to create synthetic multigrade oil with tremendous low temperature performance. Such formulations as 0W-40 and 5W-50 are now available. Manufacturers have even produced these oils without using 100 percent synthetic base-stock; such oils are called semisynthetic or part synthetic.

But can oil that is made to pump in an engine at −48° C also protect an engine at very high temperatures? The short answer is yes, and here is why. All motor oils and, in fact, all hydraulic oils have a viscosity measurement. The measure of an oil's ability to flow is its kinematic viscosity, which is expressed in centistokes (cSt) measured at either 40° C or 100° C. It is derived from the measurement of the time it takes for a fixed volume of oil to flow through a standard capillary tube. The oil company's product information handbook (available for your asking from the oil company) will list these numbers for any reputable oil. The numbers used in the following example come from the Imperial Oil, Toronto, Canada, product information handbook. The numbers from other companies will be much the same.

Example: Let's start with 15W-40 XD-3 Extra diesel motor oil; the American Petroleum Institute rates this oil CH-4 for severe service applications with turbocharged engines. The Imperial Oil Handbook says that this oil's kinematic viscosity is 118 cSt at 40° C and 15 cSt at 100° C. These numbers can be plotted on an oil viscosity logarithmic chart (Figure 5-1). The viscosity numbers are on the vertical axis, and the temperatures are on the horizontal axis. When you join the plotted kinematic viscosity numbers, you will get a line that runs from the upper left corner of the chart to the lower right corner. Thus, for any given temperature it is possible to visualize the oil's kinematic viscosity and, hence, how well it will flow. Now plot the lines for the same oil, but this time use the 0W-40 grade. At 40° C use 80 cSt, and at 100° C use 15.6 cSt. You will notice that the gradient of this line is less than the gradient for the 15W-40 oil, and the two lines intersect just above the 15 cSt mark. The ultimate multigrade oil would make a horizontal line on this chart with a constant viscosity at all temperatures.

What now becomes apparent is that the 0W-40 oil is less viscous (flows easier) at low temperatures and is more viscous (provides better protection) at high temperatures than the 15W-40 weight oil. This is contrary to what many people think when they see the low number for the 0W-40 weight oil; they assume that it is an oil for extreme cold temperature use only. The chart indicates the difference in viscosity of the two oils at 150° C is greater than one point, which can make a significant difference in the wear of an older vehicle's engine components. What makes 0W-40 oil thicker than the 15W-40 oil at high temperatures? The polymers chaining together. Thus, the same technology that makes the oil flow so well at cold temperatures makes it protect the engine parts at high temperatures.

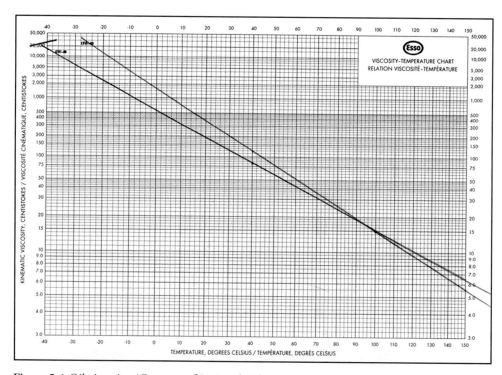

Figure 5-1 Oil viscosity. (Courtesy of Imperial Oil, Toronto, Canada)

Lubrication and Cooling System Maintenance

Grease

Grease is usually understood as a shot of something, sometimes, somewhere. However, this freewheeling attitude is a far cry from the technical knowledge and stringent record keeping that form the basis of an effective greasing regimen.

Grease vs. Oil

The first thing you need to understand is why grease is used in some applications instead of oil. The four most common criteria used to determine where grease should be used are:

- The speed of the bearing
- The load placed on the bearing
- The cooling effect required
- The location of the bearing seal

The simple diagram in Figure 5-2 displays the relationship between speed and the load on a bearing and whether grease or oil is needed. If the bearings on the shaft turn fast, oil is used. If the bearings rotate slowly, grease is used. If the load is light, an oil lubricant is used, and if the load is heavy, grease is used. Grease is needed for extremely high loads where an oil lubricant would fail. Grease is also used in sealed-for-life applications.

Oil works as a coolant removing heat from the bearings or housings. Grease is used when oil circulation is impractical, as with open gears on aerial trucks. When the bearing is a type that cannot be sealed or enclosed, then the lubricant must be thickened to stay in place. This is another situation in which only grease can be used.

Example: Let's look at a very common lubrication problem, the front-wheel bearings on a heavy truck. These bearings are usually lubricated with oil. In comparison, the front-wheel bearings on a rear-wheel-drive car are lubricated by grease. At first glance that may be the reverse of what you would expect if grease should be used as the load increases and oil should be used as the speed increases. Surely a heavy truck carrying heavy loads must put a greater load on its bearings than a car. Also, it would seem logical that a larger truck wheel would rotate slower than a small car wheel to go the same distance, and therefore would need grease, not oil.

Have the wrong lubricants been used in these applications? The truck bearing is larger than a car bearing, a lot larger, and this has two effects. The first is that there is more bearing over which to spread the truck's larger load, so in fact the load is less on each roller element in the truck bearing than in the car bearing. This means an oil lubricant can be used. The truck bearing is larger than the car bearing not only in total size but also in diameter. This larger diameter means its circumference is also much

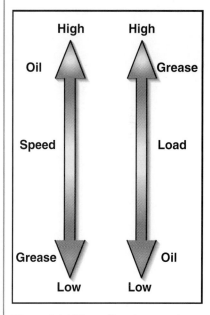

Figure 5-2 Effect of bearing speed and load on the use of grease or oil for lubrication.

Figure 5-3 Cutaway new-style wheel bearing. This new style wheel bearing has a different adjustment method from previous bearings; it is imperative that you can recognize this new design and maintain it to manufacturer's specifications.

larger, and therefore its surface speed is faster than the car bearing's, even though the large truck wheel turns slower than the car wheel. This faster speed dictates using oil to lubricate the bearing. The oil is contained in the cavity between the two bearings shown in Figure 5-3. The oil level in the cavity can be observed through a plastic window on the outboard side of the wheel.

The Composition of Grease

Emergency vehicle technicians also must understand the compositions and uses of the various grades and types of grease. Greases are classified according to their thickeners, their additives, and their grades. Normal greases start out with base oil. Thickeners and other additives are added to give the base oil certain desirable characteristics. A typical grease formula is 75 percent to 96 percent base oil (also called the fluid), 4 percent to 20 percent thickener, and 0 percent to 8 percent additive. The fluid may be a synthetic or conventional mineral oil. As the oil is the lubricant, its properties are very important when selecting a grease.

Thickeners The thickeners in grease can be metallic soaps, bentonite clays, polyurea, or inorganic compounds. The most common of the metallic soaps are calcium, aluminum, sodium, barium, and lithium. Greases formed with these thickeners can withstand far higher temperatures than those made with fats from animals or vegetables. The greases made from base oils and metallic soaps are called complex greases. They can withstand temperatures as high as 500° F (260° C). At temperatures higher than that, the lubricant will separate from the thickener. Both aluminum and lithium perform very well at high temperatures, with the lithium 12-hydroxystearate grease being the most popular.

Grades The grade of a grease is a measure of its viscosity, a very important factor in selecting a grease. Grease with too low a viscosity (too thin) will flow very easily but will not keep the metal parts from rubbing each other. Grease with too high a viscosity (too thick) will not flow in between the bearing parts and will form a channel. Some manufacturers refer to this property of a grease as its penetration or hardness. The ASTM D217 test measures this hardness. A cone of a certain weight and design is dropped into the grease; how far it penetrates will reflect the grease's hardness. A National Lubricating Grease Institute (NLGI) grade 000 grease is very thin; a grade 6 grease is very thick (Table 5-1).

Grades of 0, 1, or 2 are very common and are often specified for fire service use. Grades of 0 and 1 are usually used for vehicles stored in outdoor conditions and for vehicles operating in colder climates. As fire equipment is stored in areas above freezing 32° F (0° C), then an NLGI grade of 2 becomes more common. Most automatic grease units now being installed on trucks use grade 0 or grade 1 grease. These automatic greasers are being used more frequently on fire trucks. If you are not sure what grade of grease to use, or if your apparatus have repeated bearing failures, you can use the following calculation to determine which grade of grease will be best. First calculate the DN factor, where DN is the product of the operating speed, N, and the mean bearing diameter, D. (This is also called the surface speed of the bearing.) The DN factor is used mainly for selecting lubricants and lubricating methods.

To calculate the DN, multiply the shaft diameter in millimeters (mm) by the revolutions per minute (rpm) of the bearing.

- If DN is less than 200,000, use an NLGI grade of 1 or 2.
- If DN is at or near 200,000, use an NLGI grade of 3.
- If DN is above 200,000, convert the grease system to an oil bath system, as no grade of grease will work well. If you use an NLGI grease number 1 or 2 on a bearing that reaches speeds above 200,000, centrifugal force will throw the grease out of the bearing.

Example: A bearing in a fire pump is a ball-bearing type that can be greased. It has a shaft diameter of 26 mm and turns at 4,200 rpm. The DN speed would be 109,200; thus, an NLGI grade of 1 or 2 would work well. Now let's suppose that the manufacturer has had many bearing failures with this particular bearing. It may seem that using a larger bearing would solve this problem. If the 26-mm bearing were replaced with a 50-mm bearing running at the same speed of 4,200 rpm, however, the new DN number would be 210,000. If the same grade of grease were used with this larger bearing, more bearings would fail, not fewer.

Grade	Penetration (in tenths of a mm)
000	445 – 475
00	400 – 430
0	335 – 385
1	310 – 340
2	265 – 295
3	220 – 250
4	175 – 205
5	130 – 160
6	85 – 115

Table 5-1 ASTM D217 Test to determine the hardness of grease.

Additives Starting with high-quality base grease, manufacturers use additives to give it special characteristics. Extreme pressure additives and antiwear additives are the most common. Sulfur, zinc, molybdenum disulfide (often called moly), graphite, and fluorocarbon powders are examples of the additives used. The applications of these various greases may seem complex, but if the information is not available from the manufacturer, you can follow this general guideline: for components that will be operating in dirt and water, such as pins for stabilizing arms, gears for turntables, and sliding contacts such as steering components, use a molybdenum additive grease with a thickener of lithium or aluminum. The moly works as an extreme pressure additive keeping these slow moving but high-pressure parts from contacting each other.

Grease Incompatibility Not all greases are interchangeable, and some may be incompatible with others (Table 5-2). When mixed with another grease, some greases react and thicken. This leads to bearing failure. Many of these greases are excellent products; they just cannot be mixed. Therefore, before you change to new grease, find out if it is compatible with your present grease.

Table 5-2 Grease compatibility table.

Thickener Type	Aluminum Complex	Barium Complex	Calcium Stearate	Calcium 12-Hydroxy	Calcium Complex	Overbased Calcium	Clay (Non-Soap)	Lithium Stearate	Lithium 12-Hydroxy	Lithium Complex	Polyurea (Conventional)	Polyurea (Lubrilife)
Aluminum Complex	X	I	I	C	I	B	I	I	I	I	I	C
Barium Complex	I	X	I	C	I	C	I	I	I	I	I	B
Calcium Stearate	I	I	X	C	I	C	C	C	B	C	I	C
Calcium 12-Hydroxy	C	C	C	X	B	B	C	C	C	C	I	C
Calcium Complex	I	I	I	B	X	I	I	B	B	C	I	I
Overbased Calcium Sulphonate	B	C	C	B	I	X	I	B	B	C	I	C
Clay (Non-Soap)	I	I	C	C	I	I	X	I	I	I	I	B
Lithium Stearate	I	I	C	C	B	B	I	X	C	C	I	C
Lithium 12-Hydroxy	I	I	B	C	B	B	I	C	X	C	I	C
Lithium Complex	I	I	C	C	C	C	I	C	C	X	I	C
Polyurea (Conventional)	I	I	I	I	I	I	I	I	I	I	X	C
Polyurea (Lubrilife)	C	B	C	C	I	C	B	C	C	C	C	X

Relative Compatibility Rating

B = Borderline   C = Compatable   I = Incampatible

Applying Grease

In many cases manufacturers encourage technicians to grease components that operate in dirty conditions until the fresh grease appears, thus ensuring that any water or contamination is flushed out. This is generally the case with U-joints. The greatest cause of failure of these components is lack of greasing. Ball-bearing lubrication is very different. These bearings are generally protected from contamination by the seal, but they are subjected to wide variations in speed, load, and temperature. The prime cause of many bearing failures is overgreasing, not lack of greasing. This may sound strange, but it's true. This is especially true of bearings that are not vented, as even a cheap grease gun can produce pressures as high as 10,000 psi, which can blow out the seals and allow water and dirt into the bearing. Also, overgreased bearings will overheat because internal friction within the grease will heat the grease and the bearing. If the temperature exceeds the grease's dropping point (the temperature, measured in a laboratory, at which a grease passes from a semisolid to a liquid state), then the oil in the grease will separate from the thickeners that hold it in place. In other words the grease melts and thins, allowing the bearing and its races to contact because the thin film of lubricant no longer separates them.

Frequency You can use the following formula to determine how often a bearing should be greased.

$$t_f = K\left[\left(\frac{14 \times 10^6}{n \times \sqrt{d}}\right) - 4d\right]$$

Where:

t = service life or relubrication time in hours

f = frequency

d = bearing bore diameter in mm

n = speed in rpm

K = factor dependent on bearing type

The factors for K are as follows:

K = 1 for spherical or tapered roller bearings.

K = 5 for cylindrical or needle roller bearings.

K = 10 for radial ball bearings.

Alternatively, you can use Table 5-3 as a handy chart.

Note that a 25-mm shaft turning at 4,000 rpm would need to be regreased after one month *of continuous use* (24 hours × 30 days = 720 hours); it may take a fire pump years to reach those numbers. So you may be greasing not only too often but also too much.

Amount The formula for determining how much grease to apply is G = 0.005 DB where G = the weight of the grease in grams, D = the bearing outside diameter in millimeters, and B = the bearing width in millimeters.

Table 5-3 Grease lubrication schedule for spherical roller bearings.

Shaft Size	Amount of Grease			Operating Speed (rpm)									
Inches	MM	IN³	CC³	500	1000	1500	2000	2200	2700	3000	3500	4000	4500
½ – 1	25	0.39	6.4	6	6	6	4	4	4	2	2	1	1
1⅛ – 1¼	30	0.47	7.7	6	6	4	4	2	2	1	1	1	1
1⁷⁄₁₆ – 1½	35	0.56	9.2	6	4	4	2	2	1	1	1	1	.5
1⅝ – 1¾	40	0.80	13.1	6	4	2	2	1	1	1	1	.5	
1¹⁵⁄₁₆ – 2	45 – 50	0.89	14.6	6	4	2	1	1	1	1	.5		
2³⁄₁₆ – 2¼	55	1.09	17.9	6	4	2	1	1	1	.5			
2⁷⁄₁₆ – 2½	60	1.30	21.3	4	2	1	1	1	.5				
2¹¹⁄₁₆ – 3	65 – 75	2.42	39.7	4	2	1	1	.5					
3³⁄₁₆ – 3½	80 – 85	3.92	64.2	4	2	1	.5						
3¹¹⁄₁₆ – 4	90 – 100	5.71	93.6	4	1	.5							
4³⁄₁₆ – 4½	110 – 115	6.50	106.5	4	1	.5							
4¹⁵⁄₁₆ – 5	125	10.00	163.9	2	1	.5							

Example: Let's take a typical bearing with an outside diameter of 50 mm and a width of 20 mm. Applying the formula G = 0.005 DB:

$$0.005 \times 50 \times 20 = 5 \text{ grams of grease.}$$

How much is 5 grams of grease? It is about one shot from a standard hand-operated grease gun.

Many bearings need to be filled with grease to between only 20 percent and 60 percent of their capacity. For most bearings, 30 percent of capacity is sufficient. Many bearings now come sealed from the factory so that grease cannot be added. The reasons for this are that most people will use the wrong grease, too much grease, or incompatible greases or will not remove all dirt from around the grease nipple and thus contaminate the new grease. The contaminated grease will work like a very efficient grinding compound, and the bearing life will be severely shortened.

Note: The job of greasing is often delegated to the most inexperienced person in the fire department or the fire department shop, when it is really a highly technical task.

Fluid Analysis

Fluid analysis can involve several measurements and generate a large amount of information. One of the basic measurements is the level of contaminants in the oil. The results of this measurement are expressed as ISO ratings.

ISO Oil Cleanliness Ratings

It is very important to be able to measure the amount of contamination in a fluid and to be able to specify the level of contamination acceptable for an engine oil or a hydraulic fluid. The old saying that "if you cannot measure it, you cannot improve it" is especially applicable to fluids. Over the years there have been a number of different means to rate or measure the degree of contamination in a fluid. The one that has become most popular is the International Standards Organization for Solid Contamination Control. For many years this system measured particles of only two sizes: those equal to or larger than 5 microns, shown as ≥5, or those equal to or greater than 15 microns, shown as ≥15. We now know, however, that even particles as small as 2 microns are very destructive to a hydraulic system. These very small particles (also called silt) have an effect similar to sandblasting on close-fitting servo and proportional hydraulic valves. These valves are found on modern aerial fire apparatus, and they work very well with electro/electronic control systems. As a result, the code has been changed to reflect particles equal to or larger than 2 microns as well as the original 5- and 15-micron particles. Table 5-4 details the relation between the number of particles and ISO range numbers. For example a range number of 16 would mean that an oil sample had between 320 and 640 particles of contamination. To go down one range number, you must reduce the number of particles by one half.

	Number of particles per ml	
Range Number	More than	Up to and including
24	80,000	160,000
23	40,000	80,000
22	20,000	40,000
21	10,000	20,000
20	5,000	10,000
19	2,500	5,000
18	1,300	2,500
17	640	1,300
16	320	640
15	160	320
14	80	160
13	40	80
12	20	40
11	10	20
10	5	10
9	2.5	5
8	1.3	2.5
7	.64	1.3

Table 5-4 ISO 4406 Chart.

A typical oil analysis report may come back with the oil rated as an 18/16/13 (Table 5-5). This would mean that every milliliter of oil contained 1,300–2,500 of the 2-micron size particles, 320–640 of the 5-micron or larger particles, and 40–80 of the 15-micron or larger particles (Figure 5-4). If you installed a filter rated as a 5-micron absolute filter and your next oil report came back as 18/20/13, something would have gone very wrong with the filter because the level of contamination would have increased from a range number of 16 (320–640 particles) to a range number of 20 (5,000–10,000 particles).

Now let's look at some common oil cleanliness requirements for typical hydraulic systems. As Table 5-6 indicates, systems using only gear pumps and gear motors can have far more contamination (19/17/14) than can systems that use the much more expensive piston pumps or piston motors (18/16/13) or servo control systems (16/14/11). If you cannot get a contamination control number from the original manufacturer, check the components of the hydraulic systems on the aerial apparatus and clean the oil to the standard shown in Table 5-6 for the component that requires

Table 5-5 Relationship among range number, micron size, and actual particle count range.

Range Number	Micron	Actual Particle Count Range (per ml)
18	2+	1,300 – 2,500
16	5+	320 – 640
13	15+	40 – 80

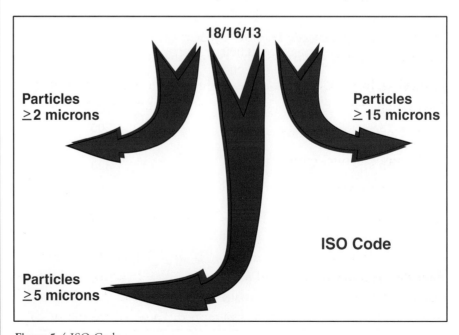

Figure 5-4 ISO Code.

Table 5-6 Fluid cleanliness required for typical hydraulic components.	
Components	**ISO Code**
Servo control valves	16/14/11
Proportional valves	17/15/12
Vane and piston pumps/motors	18/16/13
Directional & pressure control valves	18/16/13
Gear pumps/motors	19/17/14
Flow control valves, cylinders	20/18/15
New unused fluid	20/18/15

the cleanest oil. You should send fresh, unused oil for analysis as even new oil can be very dirty and may need to be filtered. Not uncommonly, new oil is in the 20/18/15 range; therefore, most manufacturers filter new oil before adding it to new machines at the factory.

Oil analysis can measure many other parameters, but the ISO ratings are the most important starting measurements for hydraulic and motor oils. These ratings and their proper use will give you an excellent starting point for selecting both the type and location of your system's filters. They can be a very useful tool anytime you have to change oil filter suppliers or have a contamination problem that you want to professionally clean up. With many modern aerial devices costing in the range of $750,000 and using upwards of 150 gallons (568 L) of oil, it is vital that you monitor oil contamination and take all steps to ensure that it meets the prescribed cleanliness standards. Remember that contamination *will* get into your system. The filter's job is to trap that contamination so it can be removed faster than it is produced. The ISO codes make it possible to judge both a filter's performance and a system's contamination.

Achieving Target Cleanliness You are likely to encounter three different filter ratings when you select a filter for either engine oil or hydraulic oil:

- Nominal filter rating

- Absolute filter rating

- Filtration ratio (beta ratio)

Nominal Rating The nominal filter rating is an arbitrary micrometer value indicated by the filter manufacturer. These ratings are very hard to reproduce outside the laboratory. A typical rating is 10 microns, which means the filter should trap particles of 10 microns or larger. In actual applications, however, not all particles of that size will be trapped. This means that most 10-micron particles would be trapped. Some 12-micron or larger particles may get through, and some particles smaller than 10 microns may be caught. This is not an accurate test of a filter's performance, but most are rated this way.

Absolute Rating The absolute filter rating is based on the diameter of the largest hard spherical (round) particle that will pass through a filter under specified test conditions. It is an indication of the largest opening in the filter elements. A typical rating may be 10 microns, which means that no 11-micron particles can pass through the filter. These filters are normally more expensive than the filters that have only a nominal rating.

Filtration Ratio The filtration ratio (also called the beta ratio) is the number of particles greater than a given size upstream of the filter divided by the number downstream. These ratios are compiled from the results of the ISO 4572 test described in the example below. This test, depicted in Figure 5-5, uses a very fine dust manufactured to a very specific size with a constant flow of oil at a constant temperature.

Example: If a filter were rated for 10 microns and the particle counter measured 5,000 particles of 10-micron dust upstream from the filter and only 1,000 particles downstream from the filter, then 5,000 divided by 1,000 would equal a beta ratio of 5. According to Table 5-7, a filter with a beta ratio of 5 has trapped 80 percent of the contamination. If the downstream numbers were reduced to 50, then 5,000 divided by 50 would equal a beta ratio of 100; this is still only 99 percent efficient. Many misinformed sales people confuse beta ratio with efficiency. It would be very easy to say that a filter has a beta of 100 and imply that its efficiency is also 100 percent; however, this is incorrect.

There is little correlation between the beta ratio and a system's cleanliness needs. A filter with a high beta ratio does remove more dirt than a filter with a low beta ratio. All reputable filter manufacturers know the ratios for their filters, but they may not want to disclose them. The machine

Table 5-7 Filtration ratio rating.	
Beta Ratio	**Efficiency %**
1	0%
2	50.00%
5	80.00%
10	90.00%
20	95.00%
75	98.70%
100	99.00%
200	99.50%
1000	99.90%
5000	99.98%

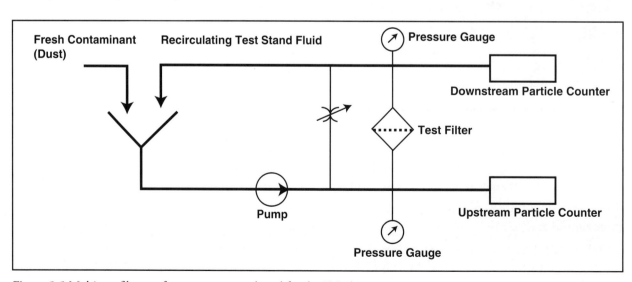

Figure 5-5 Multipass filter performance test stand used for the ISO 4572 test.

manufacturer knows how clean the oil needs to be to ensure long life of the machine. No doubt the filter manufacturer also knows this, but many are very reluctant to publish these specifications. These manufacturers say the ISO 4572 test does not relate to real world conditions, and to some extent they are correct. During the ISO 4572 test, the oil temperature and flow are held constant; in the real world, the oil temperature changes and the filter has to handle surges in oil flow as the engine speeds up and slows down. These surges can cause the pleats in the filter to bunch up and then relax. One manufacturer has overcome this by wrapping the pleats with fiberglass wrap and resin beads (Figure 5-6). Despite these variances from real-world situations, ISO 4572 still provides a useful comparison of one filter to another.

Another simple method to determine a used filter's effectiveness is to cut off the metal housing, cut open the filter, and examine the trapped material. To ensure that the cutting process does not contaminate the filter with metal filings from the outer case, use a specially designed filter cutter (Figure 5-7). If you cannot convince your department to use an oil analysis program, at least buy a filter cutter and cut open your used filters to check if the filter is trapping any major engine particles.

Figure 5-6 Filter pleats wrapped with fiberglass to prevent bunching.

Figure 5-7 Filter cutter used to remove metal filter housing for inspecting the filter.

Benefits of Fluid Analysis

You should perform fluid analysis (FA) on your emergency response vehicles for four reasons:

1. To establish a proactive (predictive) maintenance program for your equipment

2. To confirm your preventive/proactive maintenance program

3. To meet warranty requirements

4. To retain trade-in/resale value

Proactive (Predictive) Maintenance A proactive (or predictive) maintenance program based on fluid analysis will allow you to detect potential problems and make needed changes or repairs before the apparatus or its equipment breaks down.

Oil Analysis to Establish Wear Trends The generally accepted practice is to change the engine oil on an hourly basis rather than on a mileage basis. But how many hours and what kind of hours? Unlike farm equipment that registers "on-hours" on a meter when the engine runs at a rated rpm, most fire apparatus register on-hours any time the engine runs for an hour, whether at idle or full rpm. Obviously an engine idling slowly in a cool ambient temperature will degrade oil at a different rate and in a different way than an engine running at midrange rpm with a light load or an engine running at high rpm in a high ambient temperature. Remember also that engines used in the fire service are often given higher horsepower ratings than the same engines used in construction or farming. Fire department engines are often operated at full throttle. Thus, their oil and filter changes should not be based on a maintenance schedule developed for engines that serve entirely different functions in other industries.

You must also ensure that your maintenance schedule fits the particular engine you are dealing with and was not developed for an older model long since retired from service. The new electronic engines used in the fire service run drier than the older designs; that is, the top ring has been moved higher on the piston. This reduces white smoke on start-up, but places a higher soot load on the oil. Also, no other engine service places as high an electrical demand on its vehicles as the fire service. Alternators in excess of 250 amps are not uncommon. At first glance, you may wonder what relevance a high-amp alternator has to fluid analysis. While the alternator (electron pump) may be belt driven on the outside of the engine, it is gear driven on the inside. These gears drive the normal loads of the engine (oil pumps, valve train, fuel pumps, and coolant pump) as well as the extra loads of large air compressors, air conditioners, and alternators found in the fire service. A failure of these gears would be catastrophic to the engine. But fluid analysis would easily detect any wear metals in the oil, and this failure could be averted.

Liner Cavitation Another problem that timely fluid analysis can detect is liner cavitation. Cavitation is the formation and instantaneous collapse (implosion) of innumerable vapor or air pockets (bubbles) in any moving liquid. When they occur next to the cylinder liner, each of these implosions erodes away a small piece of the cylinder liner. If this is allowed to continue, small holes will be bored through the thick liner (Figure 5-8). At first, the holes will allow engine oil into the engine's cooling system, and eventually coolant (anti-freeze) will flow into the cylinder, mixing with the engine oil and leading to the destruction of the engine bearings and complete engine failure. Engine coolant not only removes vast amounts of heat from the engine, it also contains additive packages to stop liner cavitation.

Figure 5-8 Liner cavitation.

What causes liner cavitation? The answer to this question lies in the fact that many maintenance schedules have not kept pace with advances in engine design. For example, when it was first introduced, the 855 CID Cummins engine put out an amazing 220 horsepower. Now that same CID engine produces well over 400 horsepower. These changes in engine design have not only placed extra loads on the engine oil but also on the coolant. While such innovations as long-life coolant, extended drain periods, and better coolant filters have vastly improved cooling systems, coolants still have a finite effective life. After a number of years in service, a coolant may still be green or blue and good for −40° C, but it may have lost its additive package long ago. The effect of this loss will be liner cavitation. Checking the engine coolant's additive package at least every six months is necessary to ensure that the liners are protected from cavitation.

The next question is how fluid analysis can detect cavitation. At first, you may think that with small pinholes in the liners caused by cavitation the antifreeze will end up in the oil pan, but this is not always so. In many cases, the first visible sign of failure will be engine oil in the coolant. If the coolant does go into the engine oil pan, the glycol in the coolant will destroy the engine bearings long before the oil appears milky. Hence the necessity for regular oil analysis. You may also wonder why, with a pressure difference of 10 to 15 psi (maintained by the radiator cap) between the coolant and the engine crankcase the coolant will not go into the oil pan first. Even though the coolant may be at 10 to 15 psi, the oil between the piston and the inside of the cylinder liners is under far greater pressure. This pressure difference will drive the oil through the small pinholes in the cylinder liners and into the coolant that surrounds them. This is very similar to filling an oillite bronze bushing with oil and squeezing the ends with your fingers. Your fingers exert enough pressure to force the oil out through the sides of the bushing.

The oil that is forced through the cavitation holes in the liners will form a small oil scum on the coolant's surface, either in the radiator top tank or in the expansion tank. Normally, oil in the radiator is a sign of an oil cooler failure, but not always. If you find an oil scum in the radiator, have both the oil and the coolant analyzed. Another option would be to take the apparatus to an automotive shop that has either a four-gas or five-gas engine analyzer. Ask to have the sniffer probe that normally goes up the exhaust pipe of a gasoline engine inserted into the top tank of the radiator. Although this may seem to be an unusual or even humorous request, you are testing for unburned hydrocarbons. Any unburned hydrocarbons in the radiator indicate either the start of head-gasket failure, liner cavitation problems, or oil-cooler failure. Or, of course, they could indicate nothing more than careless operators who do not use clean containers when adding coolant to top up the radiator. Whatever the cause, you need to determine the problem and correct it.

Electric Currents in Engine Coolants If the analysis of the engine coolant indicates a problem with engine wear, careful maintenance of the cooling system may not be sufficient to prevent engine failure. Electrical current running through the engine coolant causes many engine failures. If an analysis of the coolant detects engine wear, it is imperative to detect and prevent any electric currents running through the coolant.

Coolant Heaters: A Solution and a Problem While starting any engine cold and immediately running it at full throttle is harmful, the nature of the fire service does not allow the luxury of a long warm-up period. In the past, the two-cycle Detroit diesel served the fire service well. This engine had a few faults, but it would tolerate cold starts better than most other engine makes. The new computer-controlled engines solve the problem of cold starts by reducing engine fuel and performance until the engine has reached operating temperature. For the fire service, however, this is not a satisfactory solution. Some departments have circumvented it by using coolant heaters (block heaters). The idea is to keep the engine up near operational temperature, which will fool the computer into giving the engine full fuel as soon as it is started. A normal block heater, however, is designed to warm a cold block so that an engine that has been outside at −20°C can be started. It was not intended to allow the engine to run at full throttle immediately after starting. In addition, plugging in a normal block heater on a hot engine at the end of a run to make sure the vehicle is ready for its next run will change the chemical composition of the coolant at the hot electrical element. To check for this chemical change, remove the radiator cap of a cool engine and smell the coolant. If it smells like fingernail polish remover, you have overheated the coolant and must change it. If you must plug in your trucks, as may well be the case in an unmanned department, use the more expensive thermostatically controlled block heaters and set the thermostats to allow engines to cool before the heaters kick in.

Water and Oil Do Not Mix Or to be more specific, water and bearings do not mix. Here is an area of maintenance that is specific to the fire service. Most fire pumps that are mounted midship are driven by a split-shaft PTO, and most of these pumps use a packing rather than a mechanical seal to seal the pump shaft. These packing seals leak. They are meant to leak; they must leak. But if the leak is excessive, the driveshaft can fling the water around and overload the seals that keep the oil in the transfer case. These transfer cases are really speed increasers, often driving the pumps at two times engine rpm. Those with gears normally use extreme pressure (EP) gear oil, and those with sprockets and chains usually use automatic transmission fluid (ATF). If even a small amount of water gets into the gear case, it will shorten the life of bearings inside. At 0.01 percent, or 100 parts per million, the bearings will serve 100 percent of their internal life. The oil will not look milky until it has reached a level of 0.10 percent, or 1,000 ppm, but by then the bearing's life will have been reduced to less than 40 percent. Long before you see the water in the oil, you will have shortened the life of the gear-box bearings. Therefore, you must analyze this oil regularly to detect any water contamination problems. What a waste of money it would be to change to an expensive new synthetic oil before you did an oil analysis, only to have water contaminate the oil.

Program Confirmation Analyze fluids whenever you change suppliers of oil or filters. This will help you determine if brand X oil is as good as brand Y or if the new filters are as good as or better than the old. Remember that oil filters differ widely in their expected performance life. Just because the purchasing agent for your city got a low price on a particular oil, do not assume that it will save money. You are going to have to prove that this oil is as good as or, hopefully, better than last year's brand. If it is not and you are stuck with a year's supply, only analysis will tell you whether to increase your oil changes. Analysis will help you to determine wear trends and adjust your oil changes accordingly. It may not be cost effective to do analysis on all the small gasoline engines in your car or light truck fleet, but sampling just a few will give you a baseline for setting oil change intervals for all of them.

Mixing brands or grades of motor oil is never recommended, but if you mix motor oils, you will at least dump most of the old during an oil change. This is not the case with hydraulic oil. Mixing brands of hydraulic oil can cause major problems because of the oils' different additive packages. Additives are used for many reasons. One is to disperse air out of the oil, and another is to remove foam from the top of the oil. Not all brands of hydraulic oils use the same chemicals to do this, and sometimes the different additives will work against one another, causing the oil to foam excessively. This could introduce air bubbles into the hydraulic oil and result in spongy controls. This may not be a big problem in a log splitter, but it could be disastrous for an aerial ladder. Any competent oil analysis company can advise you if brand X will mix with brand Y. If you switch brands, send in a sample of the new unused oil for analysis. If the oil comes in a 50-gallon (205 L) drum, roll the drum around on the floor for 10 minutes before you take a sample. You may find that this new oil has a higher-than-acceptable level of contamination and requires filtering before it can be used. This is because the large drums are reusable and may not have been cleaned as thoroughly as would be desired before they were refilled; therefore, purchasing oil in large drums may not offer the financial savings you think. Changing hand pumps from one barrel to another or using dirty buckets or pails containing everything from dead bugs to shop-rag lint will also contaminate the oil before it ever gets to the apparatus. Unless you can be sure that you can deliver the oil clean from a barrel, buy your oil in 5-gallon (20 L) pails or 1-gallon (4 L) jugs. Do not use oil that is over two years old; oil does have a shelf life.

How sensitive is fluid analysis? Perhaps a personal anecdote can answer this question best. Many years ago I was changing oil in a Liebher track backhoe (about a three-yard machine), and I wanted to sample the motor oil. I removed the drain plug just after I had shut down the hot engine. I intended to take a sample of hot oil midstream — that is to say, not at the start or end of the flow. The phone rang and I was called away on another job for two hours. Upon my return I took a sample of the oil from the cut-off steel drum into which I had drained the oil. I sent the oil away by courier that afternoon and the next day received a blistering fax and phone call from the analysis company accusing me of playing a practical joke. It seemed they had detected glycol in the oil, but a Liebher backhoe uses an air-cooled Deutz diesel. I had used that same steel barrel the week before for a radiator job on a large truck. Even though I had steamed out the barrel and drained it for a week, there was still enough glycol in its seams and pores to red flag my sample.

Warranty Requirements An oil analysis record is like a trail of footprints showing exactly what maintenance steps technicians have taken to ensure the effective operation of a vehicle. Without this paper trail there is no proof that proper maintenance has been carried out, and the warranty is null and void. The best illustration of this point is another personal anecdote.

Recently a local municipality asked me to help solve a complaint related to unusual transmission noise and a harsh gearshift on two large road graders that were just off the warranty periods. I talked to the operators, listened to the machines, and took them for a drive. I then asked the maintenance foreman for the oil analysis reports.

He said, "I do not believe in that."

I replied, "I do not believe in you."

The taxpayers of that municipality had to pay $50,000 for new transmissions because there was no trail of oil analysis footprints. Needless to say, that shop now does oil analysis.

Trade-In/Resale Value If the first three reasons for performing oil analysis have not convinced you of its cost effectiveness, this one should. Here is where fluid analysis can really pay off.

Example: Say that you have a 50–80-foot (15–25-m) aerial truck, but your city has grown, and you need a 100-foot (30-m) aerial. If your city sells or trades in the existing truck, you can demand top dollar if you have the fluid analysis records. These records can prove that you are disposing of this apparatus not because it needs a new engine, hydraulic system, or transmission but simply because it no longer meets your needs. Not only will you have the records of the maintenance that was done in the department, but you will also have reports from a third party to prove the maintenance was done. If you dispose of your equipment every ten to fifteen years, smaller departments will want to be on your waiting list.

Oil analysis is to engines what brushing is to teeth. My dentist once told me I had to brush all my teeth. I said, "All?" meaning, had I missed some? He laughed and said I did not have to brush all my teeth, just the ones I wanted to keep. You do not have to do fluid analysis on all your engines, transmissions, differentials, hydraulic systems, and split-shaft PTOs — just the ones you want to keep.

Fluid analysis is not a crystal ball that will allow you to look into the future and foresee all possible failures of your apparatus. It is, however, a very effective tool in a proactive maintenance program.

How to Read an Oil Analysis Report

Once your department has decided to contract the services of an oil analysis company, you must be prepared to interpret the results and make (or at least recommend) maintenance and service decisions. This process is best described by examining an actual case study of a 3406b Caterpillar engine.

Keep in mind the four main reasons to conduct an oil or fluid analysis program for fire apparatus.

1. To establish a proactive (predictive) maintenance program for your equipment

2. To confirm your preventive/proactive maintenance program

3. To meet warranty requirements

4. To retain trade-in/resale value

At the conclusion of this case study, try to determine if all these objectives have been met.

The department in this case study wanted to know if it could extend the oil change period and if the 15W-40 grade oil it was using was meeting the engine's needs. The department had been changing oil as close as possible to every 200 hours. This 200-hour change interval was not based on any technical information; it was just the traditional oil-change interval. The management was split over whether oil analysis was useful or a waste of time and money. While the cost of the filter and oil was high enough and the time it took to change the oil was also expensive, the greatest expense was incurred in moving the trucks to replace those that were in for service. Doing so required at least one driver and, in some cases, whole crews. This did not include the time needed to move hoses and fire-fighting equipment to the replacement trucks. If the oil change interval could be increased by 100 hours with no engine damage, then the time savings alone would make oil analysis cost effective.

The city had just changed to a new oil and filter tender, and the lowest bidder won. The mechanics had reservations about the brand of oil, and the filter was so new on the market it was unheard of. The city wanted to farm out this service work to private commercial mechanics. The firefighters were dead set against it; they trusted their own mechanics to do the maintenance work and therefore trusted the trucks to perform. During the test period, the department used the new brand and grade of oil and new oil filters purchased by the city, and the department mechanics completed the service work.

The department also was going to investigate the possibility of using an air filter dry cleaner service in the future and wanted to find a baseline for its fleet before beginning to use this new service. The department decided to buy an oil analysis service that would provide data on spectrographic analysis, physical properties, wear control, oil degradation, ISO cleanliness, and ferrography. Samples were taken over a 17-month period from 1991 to 1993. The first sample was taken during an oil change on November 7, 1991; at that time the engine had been in service for 1,395 hours, and the oil had 240 hours on it. Figure 5-9 shows the results of all the services conducted on this engine over the time of the case study. Before continuing further, let's examine each of these services and what it measures.

> **Note:** The samples for oil analysis must be taken correctly and the information supplied to the analysis company (engine hours, oil brand, grade of oil) must be accurate.

Spectographic Analysis Spectrographic analysis measures the number of different metallic particles in the oil in parts per million. It is very accurate and gets its results from burning the oil in a very controlled flame and measuring the light given off. As each metal burns, it produces a characteristic emission. You may have done something similar during a camping trip where you put a copper pipe in a fire and saw the color of the flames change. By measuring the strength of these emissions, it is possible

UNIT DATA

SAMPL #	DATE SAMPLED	COMPONENT HOURS	OIL HOURS	OIL CHNG	SMPL #
900500	03/10/93	3130.0	242.0	Y	900500
900499	10/20/92	2878.0	75.0	Y	900499
900498	09/02/92	2806.0	185.0	Y	900498
900497	07/16/92	2621.0	312.0	Y	900497
900496	05/26/92	2309.0	270.0	Y	900496
900495	03/28/92	2039.0	336.0	Y	900495
900494	01/07/92	1703.0	308.0	Y	900494
900601	11/07/91	1395.0	240.0	Y	900601

PHYSICAL PROPERTIES

GLYCOL	H2O	% FUEL	VISCOSITY 40°C	VISCOSITY 100°C	% SOLIDS
N	N	0.0	116.0	14.6	0.3
N	N	0.0	115.2	14.2	0.2
Y*	Y*	0.0	123.6	14.9	3.0*
Y*	Y*	0.0	114.4	14.1	1.2*
N	N	0.0	112.0	14.9	0.3
N	N	0.0	111.0	14.8	0.5
N	N	3.0*	61.9*	8.9*	0.4
N	N	0.0	110.0	14.0	0.2

ISO CLEANLINESS

SMPL #	3	5	15	25	ISO CODES
900500	168	58	1		13/4
900499	435	150	4	1	14/9
900498	3149	1086	13	4	17/11
900497	5997	2068	1099	379	18/17
900496	6006	2071	1190	410	18/17
900495	760	262	50	17	5/13
900494	3164	1091	174	60	17/15
900601	4060	1555	318	110	18/15

SPECTROGRAPHIC ANALYSIS (PPM)

Al Aluminum	Cr Chromium	Cu Copper	Fe Iron	Sn Tin	Pb Lead	Si Silicon	Mo Molybdenum	SMPL #	Ni Nickel	Ag Silver	K Potassium	Na Sodium	B Boron	Ba Barium	Ca Calcium	Mg Magnesium	Mn Manganese	P Phosphorus	Zn Zinc
4	2	2	38	1	2	8	1	900500	0	0	0	9	2	2	809	2	1	1200	1319
2	1	5	15	1	3	6	1	900499	0	0	0	18	4	2	837	3	2	1215	1328
9	4	211*	63	4	28*	37*	2	900498	0	0	0	346*	66*	5	875	20	3	1103	1267
7	3	74*	52	4	16*	19*	2	900497	0	0	0	126*	19	4	815	12	2	1192	1266
5	2	36*	42	2	10	8	1	900496	0	0	0	24	3	3	803	5	3	1197	1283
4	2	10	43	2	4	9	1	900495	0	0	0	7	1	2	798	5	2	1168	1278
6	3	5	55	3	7	9	1	900494	0	0	0	6	1	1	512	4	2	901	1019
3	1	2	37	1	2	7	1	900601	0	0	0	5	3	3	850	5	2	1222	1302

OIL DEGRADATION

% maximum allowable		NOX	COX	SO4	ZDDP	TBN	TAN
SOOT	OXD						
21	34	28	29	25	30	7.2	
17	14	11	15	19	16	8.5	
43	98*	95*	98*	65	85*	3.0*	
22	54	38	66	29	53	5.2	
26	33	42	43	29	38	6.4	
26.	26	38	35	25	30	6.4	
29	22	30	38	42	29	4.1	
36	21	32	20	18	35	7.4	

WEAR CONTROL CHART

Bar values: 33, 57, 175, 356, 105, 74, 88, 53 (scale 0–420)

COMMENTS: THE METAL LEVELS AND THE OIL CONDITION IS WITHIN ACCEPTABLE LIMITS FOR 242 OIL HOURS. NO COOLANT OR FUEL DETECTED. CONTINUE REGULAR MONITORING OF THE UNIT.
NEW OIL VISCOSITIES AT 40C AND 100C ARE 115cSt AND 14.3 cSt RESPECTIVELY.

THANK YOU TO: AGAT LABORATORIES
OIL PREVENTIVE MAINTENANCE
3650 – 21ST STREET N.E.
CALGARY, ALBERTA T2E 6V6

Figure 5-9 Summary of oil analyses performed on an engine over the course of 17 months.

to determine the concentrations of the metals in parts per million (ppm). Unfortunately, this method cannot detect particles much larger than 8 microns; these particles are just too large to burn and give off the correct emission and are therefore invisible to this detection technique. That does not mean this method is not useful; it just means that if this were the only method used, you could have a contamination problem involving large destructive particles that could not be detected.

The following list describes where particular metals are found in the engine. However, it is by no means complete and definitive, as engine manufacturers continue to use new materials with exotic alloys. This list should be updated regularly as manufacturers supply new information.

- Aluminum (Al): blower, camshaft bearings, turbocharger bearings, crankshaft bearings, pistons, oil-pump bushings

- Zinc (Zn): antiwear additive, oxidation and corrosion inhibitor

- Phosphorous (P): antiwear additive, extreme pressure additive

- Boron (B): coolant additive, lubrication oil additive

- Calcium (Ca): lubrication oil detergent additive, road salt, hard water

- Copper (Cu): oil cooler, valve-train bushings, camshaft bushings, thrust washers, wrist-pin bushings

- Chromium (Cr): piston rings, tapered or roller bearings, exhaust valves, liners

- Iron (Fe): cylinder liners, pistons, camshafts, crankshafts, rocker arms, valve bridges, camshaft followers

- Silver (Ag): bearings, solder

- Tin (Sn): overlay on some pistons and bearings, bushings

- Lead (Pb): bearings, solder, octane improver (gasoline engines), oil additive

- Silicon (Si): dirt, coolant, gasket sealant

- Sodium (Na): antifreeze, oil additive, road salts

- Molybdenum (Mo): coating on some engine parts, friction modifier, antiwear additive

- Nickel (Ni): bearings, stainless steel, and turbocharger blades

- Potassium (K): coolant additive

- Barium (Ba): oil or grease additive

- Magnesium (Mg): oil additive in aluminum alloys, hard water

- Manganese (Mn): oil additive, unleaded gasoline

Lubrication and Cooling System Maintenance

The sample taken on November 7, 1991, contained 3 ppm aluminum, 1 ppm chromium, 2 ppm copper, 37 ppm iron, 1 ppm tin, 2 ppm lead, and 7 ppm silicon. These first seven elements totaled 53 ppm and were used to create a graphic representation on the wear control chart. While the other metals and chemicals are important, they are not so much indicative of a wear problem as of a contamination problem or of an additive to increase oil's performance. As different oil companies will use different additive formulations, it is very important to send in a sample and establish a baseline for the new oil's additives any time you change oil suppliers.

Wear Control Chart The wear control chart gives a graphical image of the seven wear metals from the spectrographic chart.

ISO Cleanliness Numbers ISO cleanliness numbers measure contaminants that are too large for spectrographic analysis to detect. As a prepared sample is passed through an orifice, a laser beam is shone on the sensor of a particle counter. The size and amount of the interruptions can be counted, and from those data a computer can generate a report indicating the size of the particles. This is an excellent method to evaluate an oil filter's performance and to compare different manufacturers' filters.

Physical Properties As its name indicates, this test looks for any changes in the physical properties of the oil. It can detect if glycol (from the antifreeze), water, fuel, or solids have contaminated the oil and if the oil's viscosity has changed. Differentiating between antifreeze and water is very important because they can come from two different sources. In coolant leaks, the oil will contain both antifreeze and water. A defective thermostat, on the other hand, could cause the engine to run cool and allow condensation (water) to form in the oil. Fuel could come from a leaking injector, a cracked fuel line, or a damaged O-ring inside the engine head.

The next measurement is the oil's viscosity. Over time, as the lighter ends of the oil evaporate and carbon (soot) contaminates the oil, the oil's viscosity should increase. Fuel in the oil can also decrease the oil's viscosity, in effect making the oil thinner and allowing more metal-to-metal contact. These viscosity measurements are taken at both 40° C and 100° C. They are compared to the measurements taken when the oil was new (another reason to tell the analysis company the grade and the brand of oil you are using and stick with it). A quart (liter) of 10W-30 or straight 40-weight oil added to a 15W-40 oil could really distort the readings. Normally the oil viscosity should increase slightly after it is used. In this case it would not be abnormal for the viscosity to increase from a new oil rating of 118 cSt at 40° C and 15 cSt at 100° C to about 120 cSt at 40° C and 17 cSt at 100° C. If the oil's viscosity change is not linear, that is to say close to the same increase at both 40° C and 100° C, the oil's polymer additive (viscosity modifier) most likely has begun to break down. In this case, the viscosity change will show up as a decrease in the 100° C measurement of approximately 5 points from a normal of 15 cSt to a viscosity of 10 cSt.

Oil Degradation As the oil is used, it begins to degrade due to the chemical changes that occur. While the oil does not wear out, it does change because of the depletion of the additive package and contamination from acids, water, and wear metals. Soot comes from the burning (combustion) of the diesel fuel, and excess soot is a good indicator that an engine is overfueling because of a plugged air filter, defective turbocharger, or possibly a lot of stop-and-go traffic. In newer electronic engines, it could indicate that a turbo boost sensor or injector is not properly atomizing the fuel. Soot is harmful because as it increases, it absorbs zinc dialkyl dithiophosphate (ZDDP), an antiwear additive used in most heavy-duty engine oils to protect heavily loaded engine components such as valve train components. (ZDDP is very hard on catalytic converters and should not be used in oils for gasoline engines; it also will be a problem in newer generation diesel engines.)

Oxidation (OXD) results when high-temperature operation overheats the oil. These high temperatures are very common in high-horsepower turbocharged engines. The hot oil mixes with oxygen and turns black. This could easily be mistaken for soot, which also blackens the oil. Nitration of the oil is indicated by the presence of nitrous oxides (NOX); high NOX levels are a sign of sludge buildup and varnish. Solids indicate unusual levels of soot or other combustion products from either oxidation, nitration, high-sulfur-content fuel, or a plugged oil filter. Carboxylate (COX) is a measure of the oil's relative depletion of acid neutralizers. High levels of sulfation, indicated by sulfated oxides (SO_4), are caused by diesel fuel and a byproduct of combustion. As already mentioned, ZDDP is an extreme wear additive and is a measure of the oil's antiwear package. TBN is the oil's total base number, a measurement that indicates the oil's ability to absorb acids caused by combustion (think of it as the antacid of the oil world).

It is very important that the analysis company knows what oil you are using as these numbers can vary widely from one oil manufacturer to another and from one part of the country to the next because of the differing levels of sulfur in diesel fuel. This gives the analysts a baseline for determining how much the oil has changed from new. Fuels with lower sulfur content will have less acid, but the move to exhaust gas recirculation could cause higher soot levels.

Ferrography Ferrography can detect particles in the range of 1 to 250 microns, but most importantly it can tell you the particles' shapes, from which you can determine the type of wear they cause and where they came from. *Normal abrasive wear* caused by rubbing two parts together without a full oil film will produce particles in the range of 1 to 5 microns as depicted in Figure 5-10. *Cutting wear,* usually caused by abrasion of work-hardened particles in the softer metal surface produces the type of particles depicted in Figure 5-11. *Rolling* or *sliding wear,* caused by scuffing from large particles and/or fatigue, is associated with gear assemblies and

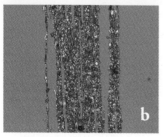

Figure 5-10 Engine particles present in the oil as a result of normal abrasive wear. (**a**) Particle type: rubbing wear; component: gearcase; metallurgy: NA; magnification: 500X; largest particle size: <15 microns; generation considered normal. (**b**) Particle type: rubbing wear; component: pump; metallurgy: NA; magnification: 500X; largest particle size: <15 microns; generation considered normal. (Both courtesy of Predict, Cleveland, Ohio)

produces the sort of particles depicted in Figure 5-12. *Rolling fatigue wear* produces either spherical or laminar particles typical of fatigue cracking in rolling element bearings (Figure 5-13).

The ferrography test is done with a microscope after the oil has been removed from the sample. Special lights and filters make it possible to determine the particle types and their form and structure. These particles could be ferrous (iron or steel) or nonferrous (aluminum, copper, lead, tin, or chrome), ferrous oxides (a sign of excessive heat, improper lubrication, or moisture content), or contamination particles (dirt, sand, dust, or other environmental particles). Although ferrography cannot tell you which bearing, cam, or piston is failing, it can tell you if wear is normal (caused by moving parts) or abnormal.

Interpreting the Results The first oil change in our case study was done on November 7, 1991; the engine had 1,395 hours and 240 hours on this oil. All samples were taken at the oil change, and the oil filter was always

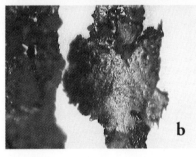

Figure 5-13 Engine particles present in the oil as a result of abnormal rolling fatigue wear. (a) Particle type: bearing wear; component: compressor; metallurgy: copper alloy; magnification: 500X; largest particle size: 70 microns. (b) Particle type: bearing wear; component: bearing; metallurgy: low-alloy steel; magnification: 100X; largest particle size: 300 microns. Generation of the particles in both a and b is due to spalling. These particles are generally very thin and may have perforations. The progression in the size and quantity of particles is the best indicator of the wear mechanism's severity. (Both courtesy of Predict, Cleveland, Ohio)

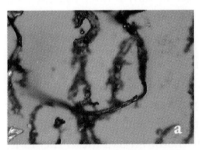

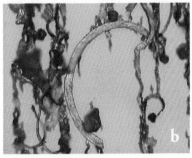

Figure 5-11 Engine particles present in the oil as a result of abnormal cutting wear. (a) Particle type: cutting; component: hydraulic; metallurgy: low-alloy steel; magnification: 500X; largest particle size: 250 microns; generation due to the presence of abrasive particles. (b) Particle type: cutting wear; component: bearing; metallurgy: low-alloy steel; magnification: 500X; largest particle size: 100 microns; generation due to the presence of abrasive particles. (Both courtesy of Predict, Cleveland, Ohio)

Figure 5-12 Engine particles present in the oil as a result of abnormal rolling or sliding wear. Generation of these particles occurs when the wear-surface stresses become excessive due to the load, speed, or reduced load-carrying capacity of the lubricant. (a) Particle type: severe sliding; component: bearing; metallurgy: white nonferrous (silver, tin, aluminum, chrome, etc.); magnification: 500X; largest particle size: 200 microns. In this case a sliding action has occurred instead of the desired rolling. (b) Particle type: severe sliding; component: worm gear; metallurgy: bronze; magnification: 500X; largest particle size: 120 microns. In this case, the worm gears had more material in contact and more sliding action generating these particles. (Both courtesy of Predict, Cleveland, Ohio)

replaced when the oil was changed. The spectrographic analysis revealed 53 ppm of the seven most important wear metals: aluminum, chromium, copper, iron, tin, lead, and silicon. All were within normal levels for oil with this many hours.

The oil was next changed on January 7, 1992, with 308 hours on this oil, an increase of 68 hours. The level of iron in the oil increased from 37 ppm to 55 ppm, probably caused by a leaky fuel injector. Diesel fuel was detected in the oil, and the oil's viscosity had dropped to nearly the rating of straight 10-weight oil. This very thin oil probably caused the higher iron levels because the oil could no longer protect the cylinder walls. This was also indicated by the increases in both chromium (piston rings) and aluminum (pistons). Interestingly, the ISO cleanliness chart revealed that the oil at this change was actually cleaner than the oil had been at the end of previous oil change periods: particles of the 3-micron size had decreased from 4,060 to 3,164; particles of the 5-micron size had decreased from 1,555 to 1,091; and particles of the 15-micron size had decreased from 318 to 174. Did this indicate a truly better filter or only a filter that worked well with very thin oil rather than with the normal 15W-40 weight that should be used?

Notice that the oil degradation chart shows the total base number (TBN) has dropped from 7.4 to only 4.1. TBN is a measure of an oil's ability to absorb acid. Where did this acid come from? The sulfur in the unburned diesel fuel. Of course, the operator/drivers who performed the daily maintenance did not notice that the motor oil had diesel fuel in it; they had been checking the dipsticks, but the diesel fuel was replacing the motor oil, so the level stayed much the same. They did notice that the oil did not appear to get as black as fast on the dipstick (as indicated by the soot levels' reduction from 36 to 29). Judging an oil's performance by looking at the oil's blackness is a mistake. A high-quality oil with high detergent levels will keep the soot in suspension and get black very fast. Soot particles are also very small and very hard to remove with normal filtration. The faulty injector was found and replaced before any further damage could be done. This should be sufficient evidence to convince the department of the efficacy of oil analysis — an engine was saved, downtime was averted.

The next change in our case study was done on March 28, 1992. With 336 hours on this oil, the wear control chart showed a decrease in metals from 88 to 74. The ISO chart also revealed a further drop; it looked as if this brand of oil filter was better than the old brand. Oil analysis is the only realistic method of comparing filter performance; anything else is just a guess.

The next change was done on May 26, 1992, at 270 hours on the oil. This change was done sooner than normal because the department traditionally experiences a very hot summer and has to deal with numerous brush fires. This prevents the department from servicing the trucks as often as it should during the fire season. The report revealed high levels of both copper and lead. Now what needed to be determined would be whether the copper and lead came from the same area of the engine or from two separate, unrelated areas. Notice the slight increase in sodium levels from 7 ppm to 24 ppm — not large enough to flag the sample but still an increase.

The next sample was taken during the height of the fire season on July 16, 1992, at 312 hours. This sample showed glycol and water (antifreeze). It was obvious now that the high levels of both copper and lead in the previous sample had come from the oil cooler. Copper is used in the tubes of the cooler, and lead is used in the solder that holds the tubes together. A tube must have come loose, and that would have been what was detected on May 26, 1992. The operators again had not noticed anything amiss in their basic fluid level checks. They had not seen a decrease in radiator levels or an increase in oil levels, and they most probably never would. The department, however, had now received two warnings about this engine and should have removed it from service, changed out the oil cooler, and removed the oil pan to inspect the crankshaft bearings. However, they decided to continue to run the engine as it was now the height of the wildfire season. Their justification was that a little antifreeze would not hurt; the engine just needed to be monitored.

On September 2, 1992, with 185 hours on this oil, the engine suffered a complete failure. This only confirmed what most mechanics already know: antifreeze and engine bearings do not mix. Notice that this oil appears very clean on the ISO chart, but the antifreeze has destroyed the engine bearings. In fact, the oil actually appears too clean to have come from an engine that has had a catastrophic failure. This engine oil normally was sampled by connecting a sampling bottle to an oil pressure passage that had a permanent sampling valve. The connection was opened, and a small amount of oil was allowed to flow out of the sampling valve to ensure the oil was representative of what was flowing in the oil galleries to the engine bearings. The sample was then taken into a clean sampling bottle. The oil that looked so clean in the sample taken after the engine failed came from the oil pan after the engine had been shut down for an hour. To be accurate the oil sample must be hot, but most importantly, it must be taken the same way each time. As you can see, the sample taken on September 2, 1992, had only four particles of the 25-micron size as opposed to the 379 taken at the previous change. The reason for the difference is simple; these big particles were at the bottom of the oil pan. Even after the engine had been turned off only one hour, most of the heavier particles had sunk to the bottom of the oil pan.

The oil degradation chart for the oil sample taken after the engine failed indicates the oil had been oxidized. Oxidation is caused by the addition of oxygen to the oil, which happens at high oil temperatures. Oil life will be shortened at temperatures above 160° F (71° C). For every 10° F above 160° F, the rate of oxidation will double with conventional oil. Synthetic oil is usually more stable at high heat but can also oxidize. Why does this oil show a high level of oxidation? The oil got super hot as the bearings on the crankshaft failed.

Luckily this engine failure did not occur during pumping when lives could have been lost. The engine was rebuilt and the truck put back into service at a cost of over $10,000. This would have paid for oil analysis on the complete fleet for the next 5 years (ten major pieces of equipment times an average of seven samples a year at $30 per sample).

The next oil change was done on October 20, 1992, at 2,878 engine hours with only 75 hours on this oil. This is normal, as any rebuilt or new engine should have its oil changed sooner than is normally recommended. Because of this early change, the wear control chart showed only 33 ppm of the wear metals.

The last oil change in our case study was done on March 10, 1993, with 242 hours on the oil. The analysis at this change showed a normal wear trend number of 57 ppm.

From this case study we can draw the following conclusions:

- Oil analysis detected the fuel leak.

- Oil analysis predicted the coolant leak by detecting the cooler failure.

- The new oil filter was doing a good job of trapping contamination.

- The new brand of oil was performing well.

- The department could extend the engine's drain period to 350 hours and gain the resulting savings.

- The recent engine rebuild was successful, and no warranty claims were needed.

What if the oil analysis had not been done? Either the new brand of oil, filter, or both might have been incorrectly blamed for the engine failure. The oil cooler would have failed anyway, even if the engine had been rebuilt after the diesel fuel leak. The annual tendering process for lubricants and filters is an outdated and costly practice that probably should be avoided. If this is not possible, the effectiveness of any new oil or filters must be proven by methods such as oil analysis. This fire department can now move on with confidence to use an air-filter cleaning service. Finally, the department now has a wear trend on which to base other maintenance directions.

Other Tests

Heat analysis is another tool that can be used to monitor the effectiveness of a vehicle's lubrication program. Whenever moving parts rub together, they create heat. This can be demonstrated simply by rubbing your hands together. This is called friction. Any moving parts in the engine that rub together produce heat that must be removed by the lubricants.

Engines also produce energy by burning fuel, which in turn produces more heat. Approximately one-third of the energy of the burning fuel is changed to heat that must be absorbed by the engine oil and the coolant fluid. Another one-third is changed to heat that is waste energy and flows out the exhaust pipe. The remaining one-third of the energy in the burning fuel is used to drive the pistons that rotate the wheels and move the truck. Any change in engine performance can be detected by both heat analysis and vibration analysis.

Heat Analysis

An infrared heat gun can detect both abnormally hot spots and abnormally cold spots in an engine. With the engine at operating temperature, point the gun at the exhaust pipes where they exit the engine (Figure 5-14). These pipes should all have approximately the same temperature. A variation of 5 percent is normal, with the front cylinders and rear cylinders often being hotter than the center cylinders. If one cylinder is too hot, the aluminum piston could very quickly melt and burn a hole through the cylinder. If the exhaust is too cold, it could indicate a compression problem (rings, valves) or an underfueling problem (maladjusted injector). This cooler temperature results when the fuel has not burned completely and released all of its energy as heat.

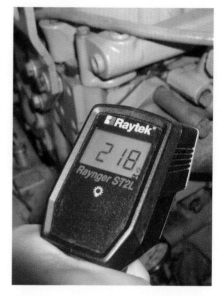

Figure 5-14 Heat analysis of an exhaust manifold on a diesel engine. Differences in temperature from one exhaust outlet to another indicate potential engine problems.

Infrared heat guns have also been used to measure the temperature of axle bearings, drive-shaft steady bearings, and oil temperature in the spilt-shaft-drive gearbox for the midship pump. Record your findings for the unit when it is operating normally as a baseline for comparison later when you may have a problem. Remember the idea is to understand the truck and prevent catastrophic problems. Detect them and solve them before the truck has a major failure.

Vibration Analysis

Any rotating component will generate a vibration, which is normal. Figure 5-15 shows a typical vibration chart for a normally operating gearbox. The almost constant vibrations shown in the figure indicate that this gearbox does not have a broken or chipped gear tooth or a worn bearing. If there were a broken tooth on a driven gear in a gearbox, it would show up with a timing that corresponded to one-half the engine's, as most gearboxes in fire trucks turn at approximately two times the engine speed. The chart in Figure 5-16 depicts a gearbox with a worn bearing. As they wear, even high-quality bearings begin to flake off material from their races. This spalling, as it is called, can be accelerated by overloading and improper lubrication. The defective bearing in the chart is developing excessive spalling marks. As the bearing fails, larger pieces of the bearing race will spall, and the bearing will begin to vibrate. This destructive vibration can cause other parts to fail, gear teeth to come out of alignment and break off, and shafts to crystallize and break.

The instrument shown in Figure 5-17 is connected to a laptop computer and produces the type of vibration charts shown in Figures 5-15 and 5-16. As with any tool, learning to use it will take time. The advantage of using a

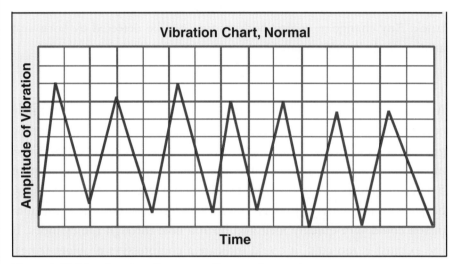

Figure 5-15 Vibration chart for a normally operating gearbox.

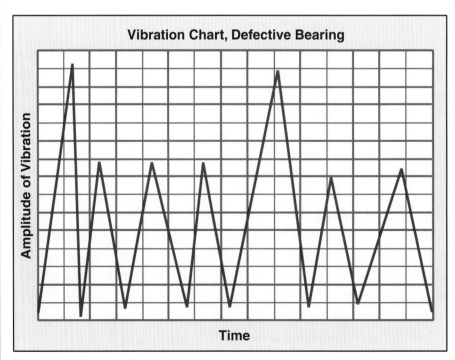

Figure 5-16 Vibration chart indicating a defective bearing in a gearbox.

Lubrication and Cooling System Maintenance

Figure 5-17 Vibration analysis on the gearbox of a front-mount pump.

computerized instrument is that the information will be stored so you can compare the vibration readings for the same components taken over time. Vibration changes can be very small and occur over a very long time. It is imperative, however, to take the measurements under the same conditions each time if the results are to be significant. For example, a vibration reading on an axle bearing under load on a gravel road will not be the same as one taken on a bearing without a load on a smooth paved road.

Cooling System Maintenance

The fire apparatus maintenance industry recently has recognized that some apparatus engine failures can be traced to problems in the cooling system. These problems have been occurring despite what would be considered sound cooling maintenance procedures. In fact, the problems have been caused not by poor maintenance of the cooling system but rather by an electrical current passing through the coolant due to electrical ground problems or to the generation of static electricity elsewhere on the equipment.

An electrical current passing through the coolant can cause the following problems:

- Pitted liners

- Failed oil coolers. Aluminum corrosion products will stop the flow of coolant through the oil cooler causing severe ring and bearing wear due to improperly cooled engine oil (Figure 5-18).

- Failed radiators (Figure 5-19)

- Extreme aluminum corrosion

Figure 5-18 Cutaway of engine oil cooler, in green, and water pump to the right, also in green. Both of these parts are susceptible to corrosion if excess voltages are not controlled.

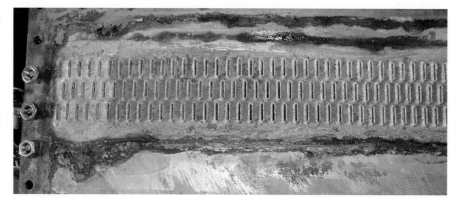

Figure 5-19 Radiator top tank and core corrosion. Note that the cores are almost closed off. This is why the fire pump service test must be conducted for the full three-hour time requirement. An engine operated for only a short time or under a light load may not reveal this problem.

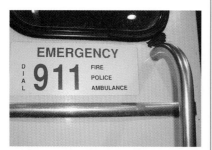

Figure 5-20 Rust has formed on the cab where the two dissimilar metals, stainless steel and regular steel, come into contact. (Source: Author)

Figure 5-21 Engine cutaway showing a copper injector sleeve. Note that this unit is not damaged.

- Abnormal water pump and head gasket failure
- Iron destruction caused by copper plating onto the iron components
- Abnormal rusting of the cab and other sections of the equipment (Figure 5-20).

Some of the documented cases that revealed the problem of electrical current in the coolant system are as follows:

- Copper injector shells in a truck engine were being destroyed in 30 days because a broken cab ground strap allowed the electrical current to ground through the coolant (Figure 5-21).
- A twelve-cylinder marine engine was destroyed by liner pitting. The overhauled engine was damaged again after only a short time. The starter, which was starting the engine with no apparent problem to the operator, was causing a 12-volt current to ground through the coolant.
- Engine blocks in a large tugboat were destroyed by pitting. Electrical switches on the after-cooler systems and one switch on the air conditioning unit in the captain's cabin caused the pitting.
- The aluminum top tanks of truck radiators were pitting on a new fleet of trucks equipped with rubber-air-bag suspension on the rear ends. The rear ends were generating a current that passed up the driveshaft to the cooling system. Grounding the rear ends and transmissions stopped the problem before the engines were destroyed.
- A large fleet of trucks made up half of fuel tankers and half of flatbeds lost sixty engines in one year. All of the trucks were of the same brand, and all used the same brand of engine. Fifty-four engines failed in the flatbeds, while only six engines failed in the tankers. Fuel tankers have a bonded ground system while flatbeds do not. This nine-to-one ratio indicated that some factor other than driver error was the cause.
- A truck hauling plastic pipe was losing the engine every 100,000 miles. The operator noted the load was glowing because of static electricity caused by the air brushing down the open-ended pipe. The operator covered the pipe with a tarp, and the engine lasted over 300,000 miles. You will notice that trucks hauling plastic pipe now most likely have the front of their loads tarped to eliminate this problem.

Electrical Test for Voltage in the Apparatus Cooling System

The only way to stop these types of failures is to correct the electrical problem causing the current. To detect this potential for engine destruction, you should incorporate a test for current flow in the coolant into any preventative maintenance program.

To conduct this test you will need a multimeter voltmeter capable of reading both AC and DC voltages. The meter must read in tenths of a volt from 0 to the maximum voltage of the system being tested. The meter leads must be long enough to reach from the coolant to the ground side of the battery. An ohm function on the multimeter is very helpful to pinpoint areas of resistance in an electrical system. These areas of resistance will cause an electrical current to ground through the coolant rather than through the engineered electrical circuit. Never connect an ohmmeter to a live circuit. The *hold, minimum,* and *maximum* functions available on higher quality meters come in very handy when taking these measurements.

Test Procedure for Mobile Apparatus:

1. Attach the proper meter lead to the ground side of the battery, negative to negative or positive to positive.

2. Install the second lead in the coolant, touching the coolant only.

3. Read the DC and AC voltage with all systems off (Figure 5-22). If a block heater is present, also take a reading with the heater turned on. If an automatic battery charger is present, as in a standby system, also take a reading with this system running. Many fire departments now use continuous-duty engine-block heaters to keep four-cycle engines up to or near operating temperature. Ensure that any 120-volt AC system is properly bonded and grounded through the use of three-prong electrical connections.

4. Read the DC and AC voltage with the electrical starter engaged (Figure 5-23).

5. Read the DC and AC voltage with the engine running and all systems turned on: lights, heaters, air conditioning, siren, fire pump primer, hose reel rewinds, and the two-way radio in both standby and transmit (Figure 5-24).

6. Remove the lead from the coolant and repeat the DC and AC voltage tests with the lead touching the outside of the engine block.

7. Keeping the lead removed from the coolant, repeat the DC and AC voltage tests with the lead touching the top radiator tank.

These procedures will test a complete system for an electrical current, which is most often generated by the rear-end differentials. This is particularly true with air-bag suspension; electrical current generated will travel up the driveshaft to ground through the engine coolant or the transmission coolant. To overcome this problem, the rear ends and transmissions should be bonded and grounding straps installed to connect the rear ends to the ground (earth).

Figure 5-22 Any voltage over 0.3 volts DC that is detected in an engine when all accessories are turned off is considered excessive for a cast iron engine. This engine will need its radiator drained, flushed, and refilled with a high-quality engine coolant after the electrical problem has been traced and eliminated.

Figure 5-23 A current of 7.86 volts DC with the starter engaged was detected on this engine. The problem was traced to a defective ground that caused the starter to ground through the radiator coolant. Note that the engine cranked over only slightly slower than normal. (Source: Author)

Figure 5-24 Currents of 4.66 volts AC (**a**) and of 1.52 volts DC (**b**) were detected in this engine when the fire pump primer was engaged. Loose and worn brushes in the fire pump primer's electrical motor caused both readings. This problem was discovered when the apparatus could not pass the annual service test because of a weak primer.

Test Procedure for Marine Apparatus:

1. Test each engine as outlined in steps one through four above.

2. Test DC and AC voltage of each engine coolant with all lights, electronics, air conditioning, and every electrical item turned on. Standby generators and main engine props should be running for this test.

3. Also, test from the outside of the engine block to the ground side of the battery.

A current of 0 to 0.3 volts is normal in the coolant of a cast iron engine. But an electric current of 0.5 volts will destroy a cast iron engine with time, and engine manufacturers are reporting that as little as 0.15 volts will destroy an aluminum engine. The current detected will be AC if the problem is due to static electricity.

Steps to Follow if an Excessive Voltage Reading is Detected

1. If the coolant shows an electrical problem with all the equipment turned on, turn off one system at a time until you finally turn off the system that stops the electrical current. Through this systematic process, you can isolate and detect the electrical system causing the problem.

2. Be particularly careful of starters, fire pump primers, and sirens. If these high-amperage accessories are generating an electrical current, they can cause as much damage to an engine as a direct connection to an arc welder.

3. Always change the coolant if an excessive electrical voltage (above 0.3 volts DC) is detected or have the coolant sent out for analysis. While the new long-life coolant is an excellent product, even it will not protect an engine with the electrical problems described above. The electrical current will destroy the iron protecting chemicals in a properly formulated coolant.

4. If aluminum damage has occurred, check the oil cooler and radiator to be sure they are not stopped up with aluminum-oxide corrosion. This can lead to liner scoring and cause engine failure.

Summary

Lubrication is a vital part of an effective maintenance program for fire service vehicles. It is a highly technical process that needs to be examined and monitored on a regular basis. The most effective tools for this monitoring process are fluid, heat, and vibration analyses and record keeping. Lubrication is an area where short-term cost cutting may not be effective for overall cost savings. The money saved by converting to cheaper oils, grease, or filters may be lost if any of the new oils or greases is incompatible with the previous lubricants or if the new filters are not as effective as the old. The bottom line, however, is not money: the bottom line is the reliability of the vehicle and the safety of those who operate it and depend upon it to do their jobs.

Chapter 5 Review Questions

Short Answer

Write your answers to the following questions in the blanks provided.

1. List the four functions of a lubricant.

   _____   _____

   _____   _____

2. What additive in motor oil allows the oil to pour at low temperatures and protects engine parts at high temperatures?

   _____

3. Name four popular grades of multigrade oils.

   _____   _____

   _____   _____

4. Will a 0W-40 oil protect an engine at high temperatures as well as a 15W-40 oil will?

   _____

5. What type of lubricant should be used to protect a bearing that turns at a very high speed, oil or grease?

   _____

6. What type of lubricant should be used to protect a bearing that has to endure extremely heavy loads, oil or grease?

   _____

7. Which grade of grease is thicker, a 00 grade or a grade 5?

   _____

8. True or false: overgreasing a bearing will cause the bearing to overheat? Explain your answer.

   _____

   _____

9. What would be the ISO 4406 test range reading numbers for an oil sample that had 11,000 particles of the 2 micron size, 700 particles of the 5 micron size, and 44 particles of the 15 micron size?

   _____

10. Would an increase of one range number double or quadruple the amount of contaminants?

    _____

11. Why are particles as small as 2 microns destructive to a high-quality hydraulic system?

    _____

    _____

12. What ISO fluid cleanliness rating is needed for a hydraulic system that uses a piston type pump?

    _____

13. Is it possible to have new oil contaminated before it even gets into the unit? Explain your answer.

_____

_____

_____

14. What three ratings will you encounter when purchasing a filter, and which rating most closely reflects a filter's ability to trap contamination?

_____

_____

_____

15. What would be the beta ratio and the efficiency rating of a filter that has an upstream particle count of 5,000 and a downstream particle count of 250?

Beta ratio _____

Efficiency rating _____

16. In reference to an oil filter, what is pleat bunching and why must it be avoided?

_____

_____

_____

_____

17. Describe how one oil filter manufacturer has avoided this bunching problem.

_____

_____

_____

18. Name the four reasons why oil analysis can be beneficial to a fire department.

_____

_____

_____

_____

19. Name at least seven wear metals found in engine oil during analysis and the components in the engine from which they might come.

_____

_____

_____

_____

_____

_____

_____

20. Name eight antiwear additives in engine oil.

_____   _____

_____   _____

_____   _____

_____   _____

21. Can spectrographic analysis measure particles in the 20 to 30 micron range?

_____

22. As engine oil becomes polluted with soot, does its viscosity increase or decease?

_____

23. What effect can long periods of idling the engine have on the engine's oil?

_____

_____

_____

_____

24. How could heat analysis be used on an engine?

_____

_____

25. How could vibration analysis be used on a gearbox?

_____

_____

_____

_____

26. List three problems caused by an electrical current passing through engine coolant.

_____

_____

_____

27. How long must the leads be on a multimeter volt meter used to test for an electric current in the coolant?

_____

_____

_____

28. To carry out a complete test of a cooling system for an electrical current, where should the multimeter volt meter leads be placed?

_____

_____

_____

_____

29. How do you prevent a current generated by the rear-end differentials from traveling up the driveshaft and grounding through the engine coolant or the transmission coolant?

_____

_____

_____

30. Describe the procedure used to detect which system is generating an electrical current in the coolant.

_____

_____

_____

31. What negative effect does an electric current have on engine coolant?

_____

_____

Chapter 6
Electrical Maintenance

Electrical Maintenance

Introduction

The fire service places unique demands on the electrical components of its vehicles. These demands must be understood when purchasing decisions are made and when predictive maintenance schedules are created. This chapter gives an overview of the unique electrical demands of fire service operation and the advantages and disadvantages of the various components available to supply those demands. Above and beyond the routine basic maintenance of electrical components described in Chapter 4 are predictive maintenance requirements for the electrical components that are either unique to the fire service or that are subject to the unique demands of fire service use. This chapter describes and analyzes these requirements.

High-Output Alternators for the Fire Service

Next to the engine and the fire pump, the electron pump (alternator) is the most important item on the truck, and great care must be taken in its selection (Figure 6-1). The wrong choice of alternators will affect the apparatus's operation and dependability for years to come. Before making a purchasing decision you have to understand the function of the alternator and the unique demands placed upon it by fire service operations.

Alternator Basics

A great deal of confusion often surrounds what an alternator actually does. Let's clarify the function of the alternator and its relationship to other electrical components.

1. Alternators do not make electricity any more than the water pump makes water. An alternator only pumps electrons that are already in the metals of the wires, frame, and batteries.

MODEL	OUTPUT
C613	14V/290A
C615	14V/340A

Figure 6-1 High-performance alternator. (Courtesy of C. E. Niehoff and Co.)

2. Voltage regulators do not regulate amperage. They regulate only voltage. If you have a lower than normal amperage output and the voltage is normal, it is not the fault of the voltage regulator.

3. A properly functioning alternator cannot work with a defective battery. Always test the battery before working on the alternator. A defective battery can make a functioning alternator appear defective.

4. Do not test an alternator by disconnecting the battery wire at the alternator when it is running and looking at the spark. This is very destructive to the alternator's components because the alternator may not be able to sense the battery voltage. The large blue spark that you see will be in the range of 150 to 200 volts or more.

5. You do not need to polarize an alternator. This notion is a holdover from the days of generators. You may need to restore the residual magnetism on some larger alternators after a rebuild or if the unit has not been used for a long time. Flashing the R terminal on an alternator that has one can usually do this. Doing this requires the engine to be off, with the ignition key in the on position. Touch one end of a wire to the battery terminal on the alternator and momentarily touch the other end of the wire to the R terminal. This should produce a small spark of approximately 4 amps.

6. Alternators will not produce their rated amperage and their rated voltage at the same time. An alternator may be rated for 250 amps and 14.4 volts, but you will never get both at the same time with a normally functioning alternator. If the regulator is functioning correctly, it will tell the alternator to cut back the amperage so that it maintains a constant voltage in the system. It may produce as much as 250 amps, plus or minus 10 percent, *or* it may produce 14.4 volts. But it will never produce both high amps and high volts at the same time. As amperes increase, voltage will decrease to as low as 12.8 to 13.0 volts. Conversely, as the voltage increases, the alternator's ampere output will decrease.

7. Alternator ampere output cannot be checked at engine idle. Most high-performance alternators must spin their rotors at least 5,000 rpm to obtain their rated ampere output. The ratio of the drive pulley (engine) to the driven pulley (alternator) causes most alternators to spin at only 2,000 shaft rpm at engine idle. This will mean a reduction in ampere output.

The chart in Figure 6-2 illustrates this. At only 2,000 shaft rpm, the model of alternator depicted in the chart will put out only 150 amps at 72° F (22° C) and 140 amps at 200° F (93° C). Notice that the alternator does not achieve full ampere output until the rotor

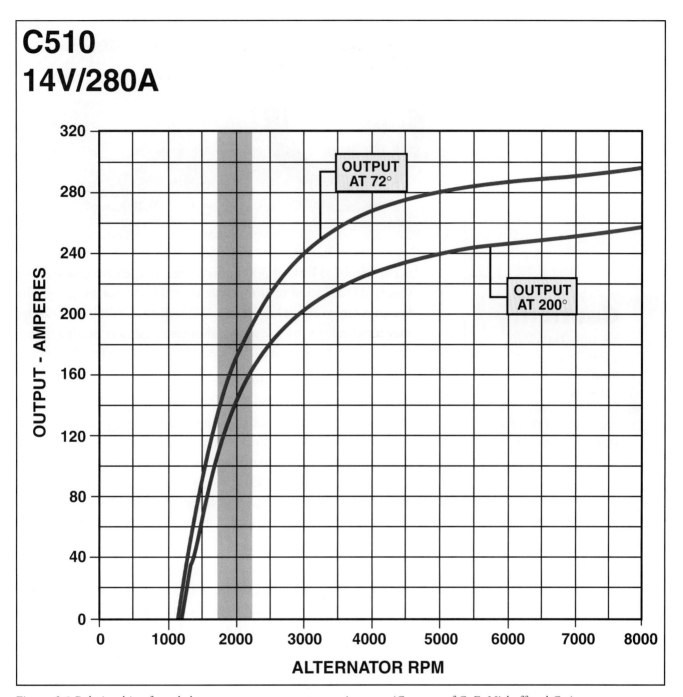

C510
14V/280A

Figure 6-2 Relationship of rated alternator ampere output to engine rpm. (Courtesy of C. E. Niehoff and Co.)

shaft is spun at 5,000 rpm. This is very common for heavy-duty alternators, and the number can be as high as 6,500 rpm for many automotive-type alternators used on ambulances.

8. Not all bearings are suitable for alternator use. The many different grades of bearings, with their different qualities of grease, will give many different lengths of serviceable life (Figure 6-3). A bearing's service life is referred to as its L 10 life. This is given in hours and is calculated when 10 percent of the test bearings have failed. This can

Figure 6-3 Cutaway views of a typical jobber bearing on the left and a factory-recommended bearing on the right. Note the extra roller bearings in the factory-recommended bearing. These roller bearings would normally be hidden behind the seals.

Caution:
Use only bearings supplied by the manufacturer. *Do not* use replacement bearings selected from a catalog of "will fits," no matter what the price.

occur within as little as 300 hours for an inexpensive bearing with standard lubrication to as much as 10,000 hours for a high-quality bearing with synthetic lubrication. While dozens of different bearings may physically fit in a fire service alternator, use only bearings specified by the manufacturer. This is no place to save a few dollars.

High-performance alternators are exposed to high heat, which destroys conventional lubrication, and to high bearing torque. Clean the diode heat sinks located in the alternator on a regular basis. This will remove any dirt or dust that will act as an insulator and trap the heat. When the alternator operates at high output but low rpm, it places a tremendous load on the rotor-shaft bearing. This occurs when the truck is at a curb with the engine at idle and when a large electrical demand such as lights and air conditioning is placed on the alternator.

Many fire department maintenance facilities send their alternators out for service on a regular basis. This is done *before the alternator fails.* Rebuilding and reconditioning can be done in the fire department facility or at a factory-authorized site.

Alternator Efficiency

The alternator's ability to convert engine horsepower into electrical horsepower is referred to as its efficiency. This electrical horsepower is measured in watts (746 watts equal one horsepower). To calculate watts, multiply the volts by the amps (Figure 6-4).

For example, an alternator that puts out 290 amps at 13 volts could supply 3,770 watts, or 3.770 kilowatts (kW), of electrical power. Many alternators that incorporate a claw pole design and brushes are typically rated at only 50 percent efficient. This would mean that for the alternator to produce the above wattage of 3.770 kW, the engine would have to put in at the belt more than 7.54 kW (10 horsepower) of mechanical energy. This can be a tremendous strain on the gear drive of the engine, another

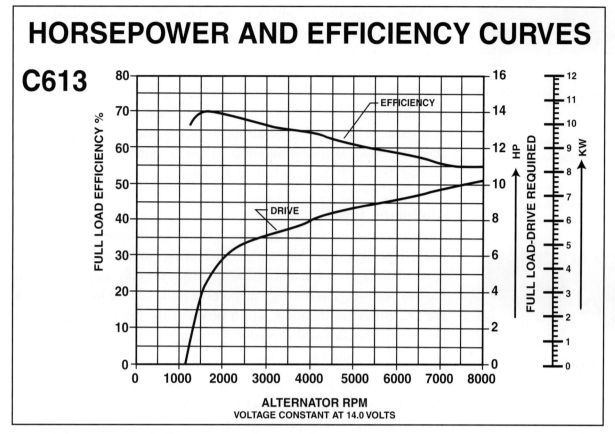

HORSEPOWER AND EFFICIENCY CURVES

C613

Figure 6-4 Horsepower and efficiency curves. (Courtesy of C. E. Niehoff and Co.)

situation where it is advisable to conduct oil analysis, as it would detect wear metals in the oil and therefore any impending gear failures.

Not all alternators use this claw pole design of rotor. Units with a laminated rotor design are capable of far greater efficiency, even as high as 70 percent. This puts less strain on the drive, uses less fuel, and prolongs bearing life. Most importantly, it generates less heat.

The mechanical energy going into the alternator from the engine can be converted into either useful electrical energy or wasteful heat energy. This heat is not only wasteful but also damages the diodes that convert (rectify) the alternating current (AC) to direct current (DC). Generally diode failure is caused by overheating. This heat is far more destructive to the diodes than welding on the fire apparatus or improper boosting.

One solution to the problem of heat buildup is an alternator design that locates the diodes outside the hot engine compartment (Figure 6-5). However, this design requires three wires from the stator winding to the rectifiers, which in turn produces more heat (Figure 6-6). DC voltage across a diode when it is in operation drops between 0.5 and 0.8 volts. This drop produces heat (voltage times amps equals watts), and this heat must be removed from the diodes. If an external rectified unit is used, you must ensure that mud does not cover the fins of the unit and block the flow of air around the diodes.

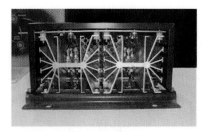

Figure 6-5 One ingenious design aimed at keeping the alternator cool is to mount the diodes outside the hot engine compartment in a large, finned, aluminum housing.

MODEL   OUTPUT
C510   14V/280A

Figure 6-6 Externally rectified brushless alternator. (Courtesy of C. E. Niehoff and Co.)

Most alternators use air to remove heat. Water-cooled diodes and engine-oil-cooled diodes have been used also, but they are no longer popular. After adjusting the belt tension, the single most important alternator maintenance procedure is to clean and remove any dirt or dust that has accumulated on the diodes and heat sinks as this accumulation prevents the circulation of air. The damage to the diodes from heat can be demonstrated by covering the air intake to the diodes at the back of the alternator. With alternator output at half of the potential amps available, the first diode will fail in as little as 10 minutes.

Alternator Types

Alternators can be divided into two groups: brush-type and brushless. Brush-type alternators have only four moving parts: two bearings and two carbon brushes. On a brushless alternator the only moving parts are the two bearings. The bearings support the rotor, and the brushes provide the DC current to the rotor, which produces the magnetic field for the alternator. The current that produces the magnetic field comes from and is controlled by the unit's voltage regulator.

All alternators have a voltage regulator; some regulators are mounted internally, some are mounted externally. Modern alternators use solid-state regulators with no moving parts. Transistors and diodes increase the field current to produce more magnetism in the rotor. This increase in rotor magnetic strength will induce an alternating current into the stator windings. This alternating current must be rectified to a direct current by the diodes. The direct current is then used to power engine, transmission, lights, and communication devices and most importantly to maintain the battery in a high state of charge. Interlocks (Appendix B) can affect the operation of these components; therefore, you need to be familiar with their location, operation, and testing.

A battery needs an electrical pressure of at least 14.4 to 14.0 volts at its terminals to fully charge. The voltage regulator's job is to maintain a system electrical pressure of 14.4 volts. If the voltage begins to drop to, say, 13.8 volts because the lights of the unit have been activated, then the voltage regulator will sense this drop and increase the current to the field (rotor). The greater field current will increase amperage output from the alternator to satisfy the loads from the lights. The voltage will return to 14.4 volts, and the regulator will maintain whatever field current is needed to maintain that electrical pressure.

As distance from the alternator (electron pump) to the battery (storage device) and therefore the length of wire can vary greatly from manufacturer to manufacturer, being able to adjust the regulator's voltage setting is often an advantage. Many regulators are adjustable, but not all; you will have to check with the alternator builder to determine this.

Electrical Problems and How to Rectify Them

In most cases if you have either too low or too high a voltage, it is the fault of the voltage regulator. Lower than normal amperage output is most likely due to a fault with the diodes or the stator windings. An open circuit in a diode or stator winding will lower the output (amperage) approximately 33 percent. A short in a diode will reduce the output approximately 66 percent and may cause the batteries to drain down when not in use. If you know what the amperage output should be, then fully load the alternator at a rotor speed of at least 5,000 rpm and record the output; if it is within 10 percent of the rated output, the alternator is functioning correctly. It is very important not to concern yourself with the voltage during this test; you are looking for amperage only.

An oscilloscope can be used to check the alternator's output. The patterns produced on the oscilloscope can indicate a failure in the diodes or stator (Figures 6-7, 6-8, and 6-9). If you do not have access to an expensive oscilloscope, you can use a voltmeter set on a low AC voltage setting (e.g., the 4-volt scale) to detect a failure in the diodes. With an alternator rotor

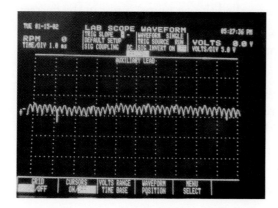

Figure 6-7 Normal diode oscilloscope pattern indicating a properly operating set of alternator diodes. Notice that the ripples are very even and regular and that the voltage difference from the highest peak to the lowest peak is not more than 0.5 volts AC. For a normally functioning alternator the voltage difference is usually 0.3 volts AC.

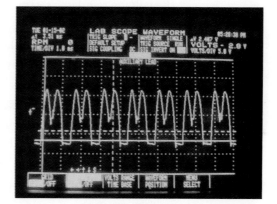

Figure 6-8 Shorted diode oscilloscope pattern. Instead of a smooth ripple pattern, the wave pattern has very deep drops. This pattern is similar to that produced by the open diode but with the added characteristic of a flattening of each negative spike. Shorted diodes get very hot and often open quickly. The voltage variation in this illustration is well in excess of 0.6 volts AC. For some shorted diodes voltage may vary as much as 2–3 volts.

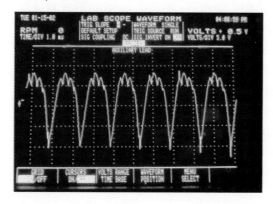

Figure 6-9 Open diode oscilloscope pattern. The voltage variation shown in this pattern is well in excess of 0.6 volts AC. For some open diodes, the voltage may vary by as much as 2–3 volts. Diodes may open from too much heat, from welding on the truck, or from incorrect boosting. Both the shorted and open diode will cause two-way radios to be noisier and may prevent the radios from receiving weak signals.

speed of at least 5,000 rpm and at least half amperage output, place the red lead of the meter on the alternator's 12-volt DC connection and place the black lead on the alternator's body. The reading should be 0.4 volts or less for a correctly functioning alternator and 0.6 volts or greater for either a shorted or open diode. Do not connect at the battery for this test as you are looking for the AC ripple caused by the defective diode and the battery can smooth this ripple giving a false reading. If the alternator cannot pass this test, then confirm the diode failure with an amperage output test.

The voltage regulator can be tested by running the alternator at a minimum of 5,000 rpm with an output load no more than 20 percent of the alternator's maximum output (for example, a 290-amp alternator should not be loaded at more than 58 amps. Slowly increasing the loads on the alternator by switching on the lights can achieve the 20 percent load. If too high a load is placed on the alternator, the voltage reading will be lower than normal and therefore false. The range of 14.0 to 14.4 volts on a cold alternator (room temperature) may decrease as much as 0.4 volts as the alternator warms up. This is common and is designed into the regulator to prevent overcharging the battery in high temperatures. If the voltage regulator setting (also called the set point) is not in the correct range of 14.0 to 14.4 volts, it can be adjusted in many cases. This could involve removing the regulator or simply removing a plate to reach an adjustment screw. Make only a small adjustment, and check to ensure that the adjusted voltage is not too high or too low. If it is too high, it will cause battery gassing as a result of overcharging, and if it is too low, it will result in an undercharged battery leading to sulfation. In the case of an external regulator, a high or low setting can be caused by corroded or loose electrical connections between the alternator, battery, and regulator. Check these connections before condemning an alternator for abnormal voltage settings.

Methods of Supplying Increased Electrical Power Demands

In recent years fire service operations have needed ever-increasing electrical power, both 110-volt AC and 12-volt DC. The 110-volt AC power must be delivered at a constant frequency of 60 cycles per second (60 Hz). A small change in frequency will not affect 110-volt lights, but it can affect computers, radios, and other frequency devices. The NFPA 1901 standard for frequency variation (Section 21-2.2) is no more than ±5 Hz. That means frequency can range from as high as 65 Hz to as low as 55 Hz (Figure 6-10). Several methods and a variety of equipment can meet this need. Each has its advantages and disadvantages, and each has unique maintenance requirements. The various options include hydraulically driven alternators, portable gas or diesel fueled generators, PTO-driven alternators, or inverters.

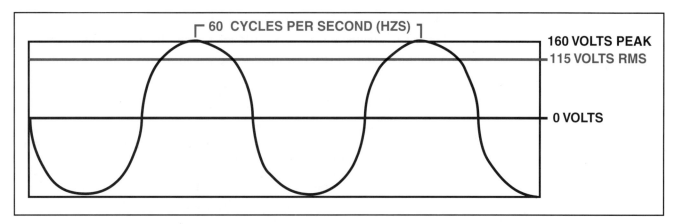

Figure 6-10 A typical 60-cycle (Hz) sine wave output from a 115-volt source. The 60 cycles are often referred to as a frequency. They are dependent on the speed (rpm) of the engine or the hydraulic motor that is turning the generator shaft. If the engine were to turn faster, then the frequency would be higher. It is very important that this frequency be held as close to 60 cycles as possible, plus or minus 5 cycles. A high-quality meter that could measure this frequency could therefore be used to set the speed of the engine or hydraulic motor.

Hydraulically Driven Alternators

Hydraulically driven alternators have many advantages (Figure 6-11). They are quiet, and in many cases they can produce their full rated output at near engine idle. Hydraulically driven generators are a costly initial investment but in the long-term are less expensive to run than portable units and weigh less. They also produce 110 volts AC far more efficiently than a 12-volt DC to 110-volt AC inverter.

The main advantage of the hydraulic generator is that it can produce a constant voltage and frequency as long as the flow of oil remains constant. This can be achieved independently of engine speed, and that means the fluctuating engine speed requirements of the fire pump do not affect the output of the hydraulic generator. The system components that keep the oil flow constant are the hydraulic pump, the hydraulic motor, and the

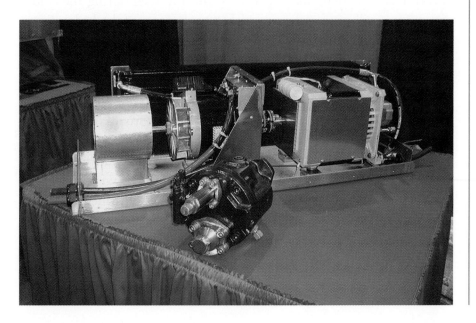

Figure 6-11 Hydraulically driven generator with, from left to right, squirrel-cage fan for oil cooler, 12-volt DC alternator, 120/240-volt AC generator, and hydraulic motor. In the background is a hydraulic oil cooler, and in the foreground is a hydraulic pump normally connected to the transmission PTO. Note the size of the red and black direct current cables for the 12-volt system.

control valves. The flow of oil determines the speed of the generator's rotor. The electrical load on the generator determines the pressure in the system. A light load means low pressure, a heavy load means high pressure.

Most high-quality hydraulic generator systems use a variable-piston pump to provide the required flow and pressures. For this variable piston pump to operate effectively and to ensure a long service life, it is necessary to maintain a supply of clean oil meeting at least the ISO level of 17/15/13 or cleaner (see Chapter 5). Some units will use both a piston-type pump and a piston-type motor to power the generator; others will use a piston-type pump and a vane-type motor. Either system is effective when used with the correct valves, but in both systems the close tolerances in the piston-type pump set the cleanliness requirements. Change the hydraulic oil every 500 hours, and take an oil sample at every hydraulic oil change to ensure cleanliness requirements are maintained.

A unique characteristic of this type of system is that the hydraulic motor, generator, and oil cooler can be located almost anywhere on the truck and can, with a little forethought, be kept away from the high under-hood temperatures that plague most engine alternators. Units with outputs of 6 to 10 kW should have a minimum of 150 square inches (968 cm²) of free, unrestricted airspace surrounding them, and units of 15 to 30 kW should have at least 300 square inches (1935 cm²). This is necessary to remove heat not only from the generator but also from the hydraulic oil.

Temperatures over 150° F (66° C) are very destructive to the hydraulic oil. The service life of conventional hydraulic oil will be shortened by 50 percent for every 10° F (6° C) the temperature rises above 160° F (71° C). Therefore, the life of an oil rated to perform for 5,000 hours will be shortened to 2,500 hours at 170° F (77° C) and to only 156 hours at 210° F (99° C). These high temperatures will cause the oil to oxidize rapidly, to turn black in color, and to smell. Varnish formed during the oxidation process will cause valves to stick. Overheating of the hydraulic oil is thus the most common maintenance problem for hydraulically driven alternators. This is because the difference between air temperature and oil temperature is quite small. For example, if the optimum temperature for the hydraulic oil were 130° F (54° C) and the surrounding air temperature were 100° F (38° C), then the Delta T (difference) would be only 30° F (16° C) — not a great deal of difference in temperature to remove heat from the oil to the air. A diesel engine, on the other hand, runs at 230° F (110° C), with a 10-pound (4.5 kg) radiator cap; if the air temperature is 100° F (38° C), then the Delta T is 130° F (72° C), which will allow a great deal of heat absorption. These hydraulic units must be placed on the apparatus where their oil coolers will be well ventilated. Most will be placed on the top of the truck.

If the unit is rated at 8 kW or more, NFPA 1901, 21-4.7.2, requires that the following instrumentation be provided at the operator's panel:

1. Voltmeter

2. Amperage meter for each leg

3. Frequency (cycle) meter

4. Power source hour meter

Examples of different gauges are shown in Figure 6-12.

Figure 6-12 Control panel for a high-performance hydraulic generator illustrating three popular styles of gauges.

PTO-Driven Generators

If the only job of the generator were to produce 14 volts DC to supplement vehicle electrical loads, then generator units could easily be driven off the transmission's power take-off (PTO). This would reduce the load placed upon the engine's accessory drives, which in many cases were never designed for 250 to 400 amp loads, and it also would remove the generator from the high temperatures under the hood. The problem arises from the need for 110-volt AC at a frequency of 60 cycles per second (60 Hz). To produce a true sine wave at 60 Hz, the rotor shaft must rotate at a constant 3,600 rpm. A small change in frequency will not affect 110-volt lights, but it can affect computers, radios, and other frequency devices. Older style units had to run at either 1,800 rpm or 3,600 rpm to produce the 60-cycle sine wave. It was not possible to produce these speeds if the apparatus were also pumping water, because the motor speed would have to be continually adjusted to meet the firefighters' water demands. With the advent of advanced electronics, manufacturers have been able to eliminate this problem.

An example of a generator employing advanced electronics to produce 110 volts or 220 volts AC is shown in Figure 6-13. These units can be belt driven from inside the engine compartment or from a PTO drive off the transmission. The unit illustrated is PTO driven. Note that the main driveshaft in the lower left of the photo goes into the fire-pump transfer case. The newer Allison automatic transmission also allows more options than before for PTO positions. This unit's unique electronics enable it to produce a very clean 60-cycle 110- or 220-volt sine wave from engine idle to full throttle and thus meet the NFPA 1901, 21-2.2, standard for frequency variation. (The unit's electronics control panel is illustrated in Figure 6-14.)

Figure 6-13 PTO-driven generator. Note the PTO shaft to the right, the drive belt and tensioner, and the aluminum generator. This particular generator may also be mounted under the hood in the engine compartment.

Figure 6-14 Gray electronic control box located behind the fire apparatus panel door. It controls the PTO-alternator shown in Figure 6-13 and ensures the correct 60-cycle frequency.

Portable-Engine-Driven Generators

Portable-engine-driven generators have been popular for many years because they are a relatively inexpensive solution to the power-supply problem (Figure 6-15). However, they are very noisy and they require a supply of gasoline or diesel fuel. These units also must be well ventilated in order to dissipate the exhaust fumes, and they always present the danger of hot exhaust burns. To prevent the possibility of explosion, they must be stopped and cooled before refueling, which is not practical in an emergency situation. They also have to be hand started or else need a battery for an

Figure 6-15 Portable engine-driven generator. Note the generator's location on a sliding drawer for easy service access.

electronic start. Finally, if the unit is fastened firmly to the truck so that it will not bounce around, then it is not very portable, and one of its apparent advantages is negated.

If your department must use a generator of this type, then a diesel unit may be the best choice. This is because the truck most likely has a diesel engine, so the one type of fuel will serve both engines. Diesel is also less flammable than gasoline, and it does not degrade as fast as gasoline in long-term storage. If the apparatus has a breathing air compressor mounted on board, then using either a gasoline or diesel unit would be unwise because of the possibility of air contamination problems.

Inverters

Another method to produce 110 volts is to use a 12-volt to 110-volt inverter (Figure 6-16). A transformer and diodes change the 12-volt direct current to a 12-volt alternating current and then to a 110-volt alternating current. The older units used a saturated iron core and did not convert power efficiently (about 50 percent, with the other 50 percent turned into

Figure 6-16 Two types of inverters: on the left an older style saturated-core inverter and on the right a new, more efficient modified sine-wave inverter. The inverter on the left can also be used to produce 12 volts DC to recharge batteries if connected to a 110-volt AC source.

heat). The last few years have seen great advances in inverter design. A switching transformer design eliminates the square-wave AC problems of the early iron-core design (Figure 6-17). These units can be used to power small, voltage sensitive 110-volt devices such as computers and computer monitors (Figure 6-18).

These new units are close to 90 percent efficient and therefore can be made smaller and do not require the massive heat-sink fins of the iron-core models. The extremely high efficiency of ferrite-core transformers is both their greatest advantage and their greatest downfall. These cores resonate at a frequency of 60 Hz and produce a harmonic wave at multiples of 60 Hz. This can interfere with the operation of two-way radios and, if they are close to the inverters, make them deaf to any incoming radio calls. Prices for units of the same output vary widely depending on shielding and design quality. If you must use an inverter, buy the best you can and get a guarantee that you can take it back if it causes radio interference. Changing 12 volts DC into 110 volts AC in order to power high-intensity floodlights is not an efficient use for a power inverter.

Figure 6-17 Square sine-wave output from an inverter.

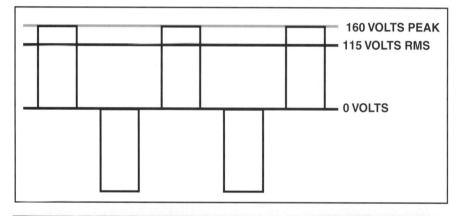

Figure 6-18 Testing an inverter. The meter on the left is able to read true RMS; the meter on the right is not and therefore shows a lower, incorrect voltage. Make sure your meter can read true RMS. Note that both of these meters read 115 volts when plugged into a standard wall socket.

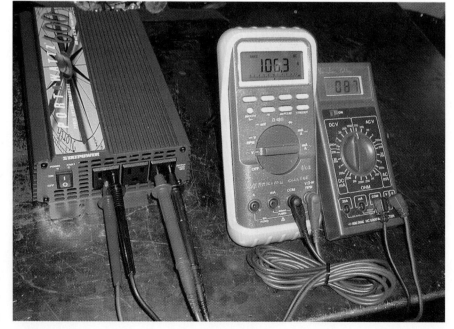

Electrical Wiring

The method for wiring an inverter to supply 110-volt AC power from the 12-volt DC batteries of the fire apparatus is shown in Figure 6-19. Use Table 6-1 to select a wire of appropriate gauge and length that will allow a drop of no more than 0.2 volts for 14-volt systems. To carry 100 amps of current at 14 volts a distance of 20 feet, you would select a wire gauge size of at least 1/0. You could use a larger wire such as a 2/0, but it would cost more. Table 6-1 also can work in reverse. That is, if you know you have a circuit that has a higher than normal voltage drop, you can conduct a circuit draw test with an ammeter and measure the length of the wire and the wire size. Then use the chart to determine if the wire is too small, the current is too high, or the wire is too long. If all of these factors are correct, then look for internal corrosion in the wire or connectors to explain the higher than normal voltage drop.

Use Table 6-2 to select wire for a 28-volt system. (As 12-volt systems are now called 14-volt systems, 24-volt systems are now often called 28-volt systems.) Why do we have 28-volt systems? In the early 1960s, 12-volt starters could not develop the torque needed to start large diesel engines. Twenty-four-volt systems could develop twice the horsepower of the 12-volt systems, as the formula for electrical horsepower shows:

$$\text{wattage} = \text{voltage} \times \text{amperage}$$

Wattage is the electrical equivalent of horsepower. If the amperage stayed the same and the voltage were doubled, then the wattage, or the power to turn the engine over, also would be doubled. As starter technology improved, 12-volt starters were able to start these larger engines.

For many years the military has used 24-volt systems, not so much to improve starting performance as to conserve material. During a war, copper becomes a strategic metal. To start the same sized engine and to

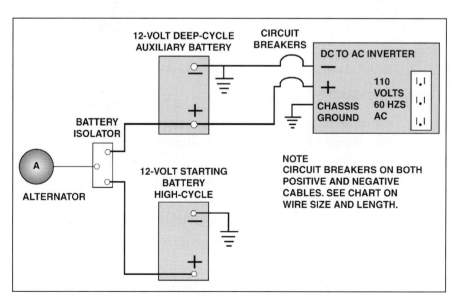

Figure 6-19 Electrical wiring for an inverter.

	12	10	8	6	4	2	1	1/0	2/0	3/0	4/0	2@2/0	2@3/0	2@4/0
30	4	6.5	10.5	16.5	26.5	42.5								
35	3.5	5.5	9	14	22.5	36.5	46							
40	3	5	7.5	12.5	20	31.5	40	50.5						
45	2.5	4	7	11	17.5	28	36.5	45	57					
50	2.5	4	6	10	16	25.5	32	40.5	51					
60	2	3	5	8	13	21	26.5	33.5	42.5	53.5				
70		2.5	4.5	7	11	18	23	29	36.5	46				
80		2.5	3.5	6	10	15.5	20	25	32	40				
90		2	3.5	5.5	8.5	14	17.5	22.5	28.5	36.5				
100		2	3	5	8	12.5	16	20	25.5	32	40.5	51	64.5	81.5
125			4	6	10	12.5	16	20.5	25.5	32.5	41	51.5	65
150			3	5	8.5	10.5	13.5	17	21.5	27	34	43	54
175			2.5	4.5	7	9	11.5	14.5	18	23	29	36.5	46
200			2.5	4	6	8	10	12.5	16	20	25.5	32	40.5
225			2	3.5	5.5	7	9	11	14	18	22.5	28.5	36
250			2	3	5	6	8	10	12.5	16	20.5	25.5	32.5
270			1.5	2.5	4.5	5.5	7.5	9.5	11.5	15	19	23.5	30
290			1.5	2.5	4	5	7	8.5	11	14	17.5	22	28
300			1.5	2.5	4	4.5	6.5	8.5	10.5	13.5	17	21.5	27
350				2	3.5	4	5.5	7	8.5	11	14	17.5	22.5
400				2	3	4	5	6	8	10	12.5	16	20
500							4	5	6	8	10	12.5	16
600								4	5	6.5	8.5	10.5	13.5

(Courtesy of C.E.Niehoff and Co.)

power the same electrical system, a 24-volt system needs less copper. Other than in the military and in airport crash trucks, 24-volt systems are rare in fire service vehicles.

Total Connected Load

One of the most misunderstood factors affecting apparatus operation is the total connected load (TCL). The relationship between TCL and alternator output is important but often overlooked both during construction of new apparatus and during the diagnosis of problems with old apparatus. The pioneering work done by C. E. Niehoff and Company has brought this deficiency to the attention of all those working in the fire apparatus manufacturing and maintenance field and must be acknowledged.

Total connected load (TCL) is the amount of electrical power needed to operate the vehicle on-scene or en route. When determining how much

Table 6-2 28-volt wire selection guide.

	12	10	8	6	4	2	1	1/0	2/0	3/0	4/0	2@2/0	2@3/0	2@4/0
30	8.4	13.3	21.2	33.7	53									
35	7.2	11.4	18	28.9	49.9									
40	6.3	10	15.9	25.3	40.2									
45	5.6	8.8	14	22.4	35.7	56.8								
50	5	8	12.5	20.2	32.1	51.1								
60	4.2	6.6	10.5	16.8	26.8	42.6	53.8							
70	3.6	5.7	9	14.4	22.9	36.5	46.1	58.1						
80	3.1	5	7.9	12.6	20.1	31.9	40.3	50.8						
90	2.8	4.4	7	11.2	17.8	28.4	35.8	45.2	57					
100	2.5	4	6.3	10.1	16	25.5	32.2	40.7	51.3					
125		3.2	5	8	12.8	20.4	25	32.5	41	51.7				
150		2.6	4.2	6.7	10.7	17	21.5	27.1	34.2	43.1	54.4			
175			3.6	5.7	9.1	14.6	18.4	23.2	29.3	36.9	46.6	58.6		
200			3.1	5	8	12.7	16.1	20.3	25.6	32.3	40.8	51.2		
225				4.4	7.15	11.3	14.3	18	22.8	28.7	36.2	45.6		
250				4	6.4	10.2	12.9	16.2	20.5	25.8	32.6	41	51.7	
270				3.7	5.9	9.4	11.9	15	19	23.9	30.2	38	47.9	
290				3.4	5.5	8.8	11.1	14	17.6	22.3	28.1	35.3	44.6	
300				3.3	5.3	8.5	10.7	13.5	17.1	21.5	27.2	34.2	43.1	54.5
350						7.3	9.2	11.6	14.6	18.4	23.3	29.3	36.9	46.6
400						6.3	8.7	10.1	12.8	16.1	20.4	25.6	32.3	40.8
500							6.4	8.1	10.2	12.9	16.6	20.5	25.8	32.6
600								6.7	8.5	10.7	13.6	17.1	21.5	27.2

(Courtesy of C.E.Niehoff and Co.)

electrical power is available, we first need to examine the batteries. The batteries *must* be, by NFPA definition, high-cycle type. Do not confuse high-cycle batteries with deep-cycle batteries, such as are used in golf carts or electrical fishing motors.

A high-cycle battery can be discharged but also provides the energy needed to restart a large diesel engine. Fire-truck batteries also must be able to deliver current over a long period after an alternator failure at the fire scene. This is why high-cycle batteries are used. For many years the batteries of choice were the D type, either 4D, 6D, or 8D, with 8D the most popular. These are not high-cycle batteries. The high-cycle battery that is most often specified now is the group 31 battery. It can supply both the high amp discharge needed for starting and the long continuous current needed in the event of low alternator output or no alternator output (failure).

The first step after determining that the battery is the high-cycle type and is fully charged is to check if the brand of alternator is the original supplied with the vehicle. The remainder of the test will provide you with the necessary data to project the on-scene survival time for any vehicle when the charging system fails. Figure 6-20 illustrates an unannotated version of this test.

> **Note:** The test to determine total connected load will be invalid if any of the batteries are defective.

Test to Determine Total Connected Load

1. A. Brand of alternator:

   At the time of delivery the manufacturer of the truck should have given the end user a nameplate displaying the rating of the alternator and the output of the alternator at both engine idle and at a 200° F (93° C) engine-compartment temperature. This information is useful to determine if the alternator is the original or a smaller replacement.

1. B. Rated charge voltage (Vc): _____

   The manufacturer should give this number as both a cold setting and a hot setting. Be careful when comparing these voltages, as a typical cold setting may be 14.4 volts DC and a hot setting 14 volts DC. This is because most alternators have some method to reduce charge voltage at this hot operating temperature. Use a digital voltmeter to take this measurement.

1. C. Maximum output rating:
   Cold (77° F [25° C])(Ac)   _____
   Hot (200° F [93° C]) (Ah)   _____

   For some brands of alternators, amperage measured hot may be 0 percent less than amperage measured cold. Get a chart of the alternator performance from either the truck builder or the alternator manufacturer (see Figure 6-2 for a typical chart). To measure the amperage, you will need an amperage gauge and a method of applying a load to the alternator. You can either use a carbon pile to do this or turn on electrical loads such as lights, sirens, and primers as long as you can apply enough loads. Most alternators must spin the rotor shaft at least 5,000 revolutions per minute (rpm) to reach full output. This requires the engine to turn at least 1,200 rpm to allow full alternator output.

   Example: In a typical setup, the alternator pulley is 2.5 inches (63.5 mm) in diameter; the engine drive pulley is 9.0 inches (228 mm) in diameter; engine full-speed maximum governed rpm is 2,100 rpm. The ratio of the drive pulley to the driven pulley is 3.6:1; for every one revolution the engine turns, the alternator will turn 3.6. For the alternator to spin at 5,000 rpm the engine must be running at 1,388 rpm.

   You will have to do this math and calculate the pulley relationships for every truck you look at. Sometimes personnel put the wrong pulley on the alternator. It might seem that the solution to the problem of low alternator output at idle would be simply to put a smaller pulley on the alternator; then it would turn faster at a lower engine rpm. However, it would also turn faster at higher engine rpm, and at the engine maximum governed speed of 2,100 rpm the alternator rotor would be turning at 7,560 rpm. If a smaller pulley were used, then the rotor shaft and its bearing could turn at far too high a number of rpm.

   ### Caution:
   Most manufacturers recommend that the rotor bearing not turn at more than 8,000 rpm for long bearing life.

Figure 6-20 Test to determine total connected load.

1. D. Amperage output at engine idle (A at idle) ————————

   Take this measurement with the engine at its normal idle speed and with an engine compartment temperature of at least 200° F (93° C).

1. E. Amperage output at 1,000 engine rpm (A at 1,000 rpm) ————————

   Take this measurement with the engine running as close to 1,000 rpm as possible and an engine-compartment temperature of at least 200° F (93° C).

2. Vehicle Engine Type

   Engine manufacturer _____

   Model _____

   Cubic-inch displacement _____

   14-volt system _____

   — or —

   28-volt system _____

3. Total Connected Load

   To determine the total connected load you can either:

   A. Assign a value to all the electrical accessories that might be on at any given time. This information should be available from the manufacturer of that component.

      — or —

   B. Take an actual on-vehicle test of all the possible loads that could be in use. This could be measured with an amperage meter connected at the batteries when the engine is off so that all power must come only from the battery packs. As some of this load will be applied only while en route and some only on-scene, you will have to take two measurements with the appropriate loads switched on or off. Consult your department's SOPs to find out what loads and lights are used in which situations.

   C. Total Component Load: Components' Amperage Draw

      (1) Dedicated loads

         (a) Engine fuel management: _____

         (b) Transmission computer: _____

         (c) Amount necessary to float charge the battery pack: _____

             If the engine senses too low a voltage at its computer, it will shut down. (Unlike the high-temperature or low-oil-pressure shutdown on fire service vehicles, this shutdown cannot be disabled.) A typical low-voltage shut down is 10.2 volts. A typical load for the engine could range from 10 to 35 amps, depending on the engine type, and for the transmission about 10 amps. To maintain the batteries at peak performance, a single 8D battery will need 15 amps, while group 31 batteries will need approximately 5 amps each.

             Total Amps Dedicated _____

      (2) Switchable loads (worst case condition)

         (a) Calculated amp value ———————— at rated test voltage (VT) ————————

             —or —

         (b) Actual amp value ———————— at battery test voltage ————————

         This test voltage will most likely be about 12.0 to 12.1 volts depending on the condition of the batteries and the size of the loads. It is measured at the battery. Be careful to take safety precautions around the batteries. To avoid any sparks, use a clamp-on type amperage meter. Here are some typical load draws:

             Lights running: 40 amps

             Lights at scene: 40 amps

             Sirens: 250 amps (mechanical), 40 amps (electronic)

             Driveline retarder (electrical type): 250 amps

Figure 6-20 Continued.

Air conditioning (fans and compressor clutch): 40 amps

Radios and cellular phones: 10 amps (receive), 35 amps (transmit)

Primer pump: 220 amps

Total Amperage Load: Switchable _____

Total Connected Load (3.A + 3.B) _____

4. Fill in the blanks to find the voltage factor (Vf):

(Vc) _____ − (Vt) _____ + (Vd) _____ = _____ + 1 + (Vf) _____ (Vt) _____

It is not enough just to put back into the electrical system what you took out to run the loads. You must also calculate the voltage factor (Vf). This is because the electrical power you used was at 12.2 volts while the power you must put back in is most likely going to be in the range of 14.0 to 14.4 volts. This voltage factor is used as a multiplier to find the total connected load. This adjusts the TCL for any given charge voltage value when the test voltages in 3.C(2)(a) (calculated amp value) and 3.C(2)(b) (actual amp value) are known. This larger number is the true value of the TCL when the system is operating at charge voltage; it represents an increase in amps of anywhere from 15 percent to 25 percent.

The voltage factor is determined by first obtaining the voltage difference (Vd) between the system's charge voltage (Vc) and the system's voltage during the load test (Vt) or, if calculated, the operating voltage at which the manufacturer rated the component lamps or devices. This voltage difference (Vd) is then divided by the test voltage (Vt). The resulting answer is the voltage factor (Vf). The product of multiplying the voltage factor by the TCL is the true amount of electrical power needed for the system.

5. True value of total connected load at charge voltage:

A. Total connected load from section 3.C: _____

B. Voltage factor (Vf): _____

Adjusted TCL (5.A × 5.B): _____

6. Output of alternator hot (Ah, from section 1) _____

Adjusted TCL from section 5 _____

Is the alternator's output when hot greater than the adjusted TCL?

YES _____ NO _____

If yes, the alternator is properly sized.

If no, then a larger alternator is needed or you will have electrical problems.

7. CID (cubic inch displacement) of engine _____ × 2 = minimum cold cranking amps (CCAs) required. (The metric calculation is 125 CCAs per liter of engine displacement.) Battery manufacturers will state the CCA rating of the battery; this measurement is always taken at 0° F (−18° C). (Check with the manufacturer. Many recommend three times the cubic inch displacement, but this is usually calculated for diesel engines that have to start outside in cold weather, which may not apply to your fire truck.)

8. Number of batteries needed to meet the reserve capacity (Rc) requirements if the alternator is off-line (failure of alternator or engine stall).

A. TCL (adjusted from line 5) _____

B. 25 (Rc load test)

8.A divided by 8.B = number of batteries: _____

Rc is the duration in minutes that a battery can deliver 25 amps and maintain a voltage above 1.75 volts per cell. As a 12-volt battery will have six cells, at the battery terminals the voltage should be not less than 10.5 volts. This is very important for the newer electronically controlled engines and transmissions because a low voltage will cause these engines and transmissions to shut down. Simply put, Rc is how long, in minutes, the truck can operate with no alternator output.

9. Total CCAs available.

A. CCA rating of battery: _____

B. Number of batteries in the pack (from item 8): _____

(9.A × 9.B) = _____ (this value must be larger than the total from line 7)

Figure 6-20 Continued.

RFI/EMI

Radio frequency interference (RFI), sometimes called electromagnetic interference (EMI), is becoming an increasingly common problem with electronically controlled transmissions, fire-pump governor controls, and other voltage-sensitive electronic devices. A magnetic field forms around any conductor (wire) carrying an electrical current. Its strength is in direct proportion to the amperage of the current passing through the wire. The higher the current, the stronger the magnetic field. If the field is created with a 60-hertz standard household alternating current (AC), then the field will build and collapse 60 times per second, or 3,600 times per minute. If the field is created by a direct current (DC), such as from an engine starter, then it will expand with the current as long as the current is increasing and collapse when the current stops flowing. If this magnetic field is strong enough, it can cause EMI and can invade the wires of sensitive electronic components, thus causing erratic operation of these devices.

This can be illustrated on a 1991 GMC truck. With the engine stopped, turn the ignition key to the run position and place a 100-watt pistol-type soldering gun within a few inches of the distributor in the engine compartment (Figure 6-21). With the soldering gun plugged into a standard household 60 Hz power source, squeeze the trigger on the soldering gun. You will notice that the two fuel injectors in the throttle body are injecting gasoline fuel. What you may not realize is that the ignition coil is also building and collapsing its magnetic field. This causes the ignition coil to discharge. If you hold the coil wire close to a steel component, it can cause a very strong spark to jump from the wire to the steel component (Figure 6-22).

Figure 6-21 Electrical soldering gun creating interference to the ignition coil on a 1991 GMC truck.

Figure 6-22 Spark (in red) jumping between the coil wire and the threaded adjuster.

If you were to measure the spark rate, you would find that the coil is discharging at the rate of 1,800 rpm. Remember that a four-cycle engine requires two revolutions to fire the spark plugs once. What causes these injectors to inject fuel and the coil to discharge a spark? The EMI from the soldering gun causes a magnetic field to invade the ignition pickup coil under the distributor tower. This EMI causes voltage to flow in the pickup coil and wires, sending a false signal (input) to the engine's computer that the engine is running at approximately 3,600 rpm. The computer then turns on the fuel injectors and fires the ignition coil (outputs).

For another example of EMI, remove the ignition-coil-to-distributor wire (approximately one foot [30 cm] long). Replace this wire with a piece of high-tension ignition wire approximately 16 feet (5 m) long. Place one end into the ignition coil tower and the other into the distributor cap. Use electrical tape to tape the wires together every two feet making sure the wires do not come into contact with the fan blades or the fan belts Cut the loop now formed at the free end. Hold the wire ends together with a pair of spark plug boot pliers. Crank the engine; it should start and run. Slowly pull the wire ends apart. There should be a large arc and the engine should keep running. As you move the wire ends apart, the spark will get longer as it tries to jump the gap.

When the wire ends are pulled too far apart, the spark will stop jumping the gap, but the engine will not stop running. Why? Because the ignition coil is building up its maximum voltage. In the 1991 GMC truck this would be 30,000 volts; in a new truck this number can approach 50,000 to 60,000 volts. (This voltage is not harmful to the demonstrator as it is the amount of current flowing, or amperage, that is lethal, not the voltage. In this situation if we use the formula $P = E \times I$, where P is in watts, E is in volts, and I is in amps, we can calculate the amount of power produced by an ignition coil. The normal coil voltage is less than 12 volts, but for this example we will use 12 volts, and the amperage is 3 amps. Substituting in the formula we have 12×3, which equals 36 watts of power. Now if the output voltage created by this 36 watts of power were 50,000 volts, then applying the formula again, the amperage or current produced would be 36 watts divided by 50,000 volts. This would equal 0.00072 amps, or 0.72 milliamps, which is less than the 1 milliamp required to produce even a slight sensation on your hand.) This high voltage is creating a magnetic field around one of the wires, and the magnetic field in turn is creating a voltage in the other wire even though the wires are not electrically connected. They only have to be close to each other for this to happen. This voltage can be at least 7,000 volts, which is enough to fire the spark plugs at engine idle.

What does this tell us about electronically controlled diesel engines, transmissions, and fire pump governors? Many of the sensors, such as those governing speed, temperature, and throttle position, generate very small voltages, from 0 to 5 volts in many cases. If they came in close proximity, the high-voltage wires could induce a false voltage into any of these sensors' wires.

Fire department personnel use two-way radios clipped to their harnesses. When these devices transmit, they radiate a signal that can produce radio frequency interference (RFI). RFI is very similar to EMI and has about the same effect on electronic components. The closer the radio is to a component when it transmits, the more voltage it will induce into the component. This can affect electronic pump governors and voltage-shedding devices. The amount of induced voltage is the inverse of the square of the distance. If you double the distance of the antenna from a component, then the amount of induced voltage will be reduced by a factor of four. This condition should be easy to detect. When the firefighter talks on the two-way radio, the engine could rev up or down and the transmission could shift up or down in turn.

A unique problem can occur with cell phone radio use. A cell phone that is turned on transmits a radio frequency (RF) signal to the cell-phone tower site. This signal can induce a voltage into sensitive electronic components and cause them to operate erratically. Because this happens randomly and without the firefighter's knowledge, problems of this nature can be very hard to diagnose.

A number of measures can help to reduce the RFI/EMI problem:

- Have any two-way radios that are mounted in the vehicle installed by a professional radio technician.

- Mount the antenna as far away as possible from any electronic devices. Ensure that only high-quality coaxial antenna cable is used and that solder joints are made correctly and are tight.

- Keep the antenna cables as short as possible and do not coil excess cable because this will induce RFI/EMI to other electronic components.

- Repair any antenna as soon as it is damaged.

- Do not alter the length of the antenna as it is critical to the operation of the radio and the reduction of RFI/EMI.

- Ensure that any wiring repairs made to the truck's electronic devices conform to the manufacturers' recommended practices.

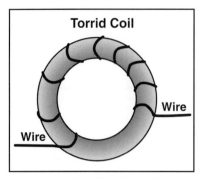

Figure 6-23 Torrid coil used to reduce unwanted induced voltage.

- If all else fails, it may be necessary to place an RF shield around the affected wires. This may be either in the form of a coaxial shield cable with a wire to a solid ground like the truck frame or in the form of a torrid coil. To make a torrid coil, wrap the wire of the affected device around a doughnut-shaped magnet (available from your local electronics store) five to ten times (Figure 6-23). Secure the wire with electrical tape. This device will reduce the unwanted induced voltage.

Summary

An EVT who is aware of the special electrical demands placed on fire-service vehicles and the NFPA standards in place to ensure that those demands are met is well placed to offer expert advice to the decision makers who purchase fire apparatus. This expert knowledge of electrical systems and electrical demands also ensures that the EVT can perform appropriate preventive maintenance, testing, and record keeping. Proactive maintenance of the most appropriate electrical components will ensure that firefighters are not left in the dark and incommunicado at the emergency scene.

Chapter 6 Review Questions

Short Answer

Write your answers to the following questions in the blanks provided.

1. An alternator could also be described as a/an _____ .

2. What do alternator regulators regulate?

   _____

   _____

3. Which terminal would you flash to restore the residual magnetism in an alternator that uses residual magnetism?

   _____

4. Should voltage be 12 volts or 14 volts at the output terminal of a properly functioning alternator with the engine at a normal idle?

   _____

5. An alternator's amperage output is considered normal if it is within what percentage of maximum?

   _____

6. What is considered the minimum alternator shaft rpm with which maximum output can be obtained?

   _____

7. As the temperature of an alternator goes up, its amperage output goes _____ .

8. High-performance alternator bearings use a long-life high-temperature _____ grease.

9. The single most important service work a technician can do to ensure long alternator life is to _____ _____ the diodes.

10. If all the energy from the engine that goes into an alternator is not converted into electrical power, what is the residual energy converted into?

    _____

11. The two types of alternators are brush type and _____ .

12. Expressed as a percentage, how much will an *open* diode reduce the output of an alternator?

    _____

13. Expressed as a percentage, how much will a *shorted* diode reduce the output of an alternator?

    _____

14. What measuring instruments can be used to detect a defective diode on an operating alternator?

    _____

15. Which type of hydraulic pump do hydraulically driven alternators use, gear or piston?

    _____

16. What minimum level of ISO oil cleanliness can be allowed for a hydraulic pump to have a long service life?

    _____

17. What four meters must be supplied for a generator of 8 kilowatts or larger?

_____   _____

_____   _____

18. Why must portable engine-driven generators be kept away from breathing air compressors?

_____

_____

_____

19. What size wire is recommended for a length of 10 feet and an amperage of 125 at 14 volts?

_____

_____

_____

20. With the correct wire size, voltage drop should not exceed _____

21. Total connected load is the amount of electrical power needed on the fire scene or _____

_____ .

22. Which are the best batteries for the fire service, high-cycle batteries or deep-cycle batteries?

_____

_____

_____

23. The reserve capacity of a battery is the amount of time in minutes the battery can deliver how many amps at what minimum voltage?

_____

_____

_____

24. Why is reserve capacity important?

_____

_____

_____

Chapter 7
Pump Testing

Pump Testing

Introduction

The vast majority of fire pumps are of the semicentrifugal nondisplacement design, which is the type discussed in this chapter. The standards for fire pumps are found in NFPA 1901, *Standard for Automotive Fire Apparatus,* and in NFPA 1911, *Standards for Service Tests of Fire Pump Systems on Fire Apparatus.*

Fire Pump Basics

Before describing any of the tests required for fire pumps, the basic principles of the fire pump will be reviewed.

Suction vs. Atmospheric Pressure

Pumps do not suck water; atmospheric pressure pushes water into the pump. Atmospheric pressure is all around us and is commonly calculated as 14.7 psi (101.2 kPa) at sea level on a normal day. This pressure is also expressed as 29.9 inches of mercury. Thus, a pressure of 1 psi (7 kPa) would push 2.035 inches (5.17 cm) of mercury up a glass tube. The pressure exerted by the atmosphere decreases as elevation increases. For every 1,000 feet (304.8 m) above sea level, the pressure decreases 1 inch (2.54 cm) of mercury or 0.5 psi (3.45 kPa). If you were to travel above the atmosphere into space, the atmospheric pressure would be near zero, an almost perfect vacuum.

If your department operates at a location greater than 2,000 feet (610 m) above sea level, the reduced atmospheric pressure begins to affect pump performance; there is just not as much atmospheric pressure to push the water up the hard suction hose and into the eye of the pump. Why does the water flow into the pump in the first place? It flows into the pump because a primer removes some of the air from the pump to create a partial vacuum. The atmospheric pressure that is pressing on the water in

Note: Mercury is no longer used to measure air pressure because it is very dangerous to both the handler and the environment, but the term *inches of mercury* has stayed with us.

the lake or drafting pit then pushes the water up the hard suction line and into the fire pump. If there were no atmospheric pressure on the water, it would never flow up the hose no matter how much you primed the pump. In other words, the fire pump could turn for hours at a very high speed and never pump water.

When atmospheric pressure is low, as on a rainy, stormy day, the effect on pump performance is twofold: the pressure to push the water up the hose is reduced, and engine performance to power the fire pump decreases. The performance of an engine depends upon its ability to get air into its cylinders. A low atmospheric pressure forces less air past the air cleaner and into the engine.

The relative humidity and the air temperature also affect an engine's performance. As the relative humidity increases, the amount of water in the atmosphere increases and thus the amount of oxygen available for combustion decreases — not to mention the obvious fact that water does not burn well. When the air temperature is high, the air expands and the oxygen molecules are farther apart; therefore, not as much oxygen is available to support combustion in the engine. As the pistons move down in their bores, atmospheric pressure will push the air into the cylinders in just the same way as water is pushed into the fire pump.

Pressure vs. Flow

Pumps do not make pressure; they only make flow. Many people have the misconception that pumps make pressure. Actually, however, pumps make flow; restrictions make pressure or, more correctly, cause pressure. In a nondisplacement pump, the amount of flow decreases as pressure increases. In fact, it is possible to totally stop all flow from the outlet of the pump for a very short time; however, this heats the water very fast, and the hot water could cause burns if a discharge were opened and the water contacted the skin (Figure 7-1). Therefore, operating the pump without a discharge opened is not recommended however briefly. Consequently, no pump performance test requires all discharge outlets to be closed. In short, don't do it.

The pressure created at the discharge outlets results from the restriction of pump flow. If all the pump discharges were opened with the pump at maximum flow, the pump panel would record a very low pressure of about 20 to 35 pounds (9.0 to 16.0 kg). Why is the pressure so low? With all the pump outlets open, the plumbing and the valves on the discharge side of the pump create the only resistance to the pump flow. As the outlets are closed off, the system pressure increases because the pump flow remains almost the same.

Figure 7-1 SMVD valve melted as a result of high heat generated when discharge outlets were closed. (Courtesy of Dave Sweden)

NFPA 1901, *Standard for Automotive Fire Apparatus,* Section 14-2.4.1, very clearly states the standard for a typical pump's suction capability. The pump must be able to pump 100 percent of its rated capacity at 150 psi (1,035 kPa) from draft when connected to a 20-foot (6-m) long suction hose with a strainer attached.

> **Note:** NFPA 1901 requires only that the hose be 20 feet (6 m) long, not that the pump must operate with a suction lift of that distance.

You can use Table 7-1 to calculate the size of the intake hoses and the maximum lift the pump should produce to pass the drafting test. The table indicates how much lift is allowed. Notice that a pump with a capacity of 1,500 gallons (5,678 L) must use both of its 6-inch drafting intakes, not just one. This is also true for a 1,750 gallon (6,624 L) pump and a 2,000 gallon (7,570 L) pump. It is not always possible to find a drafting pit that will allow access to both sides of the pumper. The solution in this case is to connect to a high-capacity hydrant. Also note that pumps of 1,750 gallons (6,624 L) or more have a reduced maximum lift height (distance from water level to the center of the pump eye).

Table 7-1 Suction hose size, number of suction lines, and lift for fire pumps.

Rated capacity gpm (l/m)	Maximum suction hose size inch (mm)	Max. number of suction lines	Maximum lift ft (m)
250 (950)	3 (76)	1	10 (3)
300 (1,136)	3 (76)	1	10 (3)
350 (1,325)	4 (100)	1	10 (3)
450 (1,700)	4 (100)	1	10 (3)
500 (1,900)	4 (100)	1	10 (3)
600 (2,270)	4 (100)	1	10 (3)
700 (2,650)	4 (100)	1	10 (3)
750 (2,850)	4.5 (113)	1	10 (3)
1,000 (3,785)	5 (125)	1	10 (3)
1,250 (4,732)	6 (150)	1	10 (3)
1,500 (5,678)	6 (150)	2	10 (3)
1,750 (6,624)	6 (150)	2	8 (2.4)
2,000 (7,570)	6 (150)	2	6 (1.8)
2,250 (8,516)	8 (200)	3	6 (1.8)
2,500 (9,463)	8 (200)	3	6 (1.8)
2,750 (10,410)	8 (200)	4	6 (1.8)
3,000 (11,356)	8 (200)	4	6 (1.8)

Source NFPA 1901.

Color Coding of Fire Hydrants

While not entirely satisfactory, individual flow tests of fire hydrants do provide approximations of the flow available from those specific hydrants. When flow tests are conducted and hydrants are marked according to some predetermined code, the ready knowledge of fire flows is of substantial value to fire and water departments. The NFPA Committee on Public Water Supplies for Fire Protection recommends four basic colors to classify fire hydrants according to flow (Table 7-2).

Available flow capacities are rated from results of a flow test that is conducted during a period of ordinary demand. The ratings should be based upon 150 kPa residual pressure. The recommended color scheme provides simplicity and consistency in a system to denote available fire flow. The barrels of fire hydrants may be any color that does not clash or distract from the color code. Chrome yellow is an accepted color in many areas. All location markers for flush hydrants should conform to the same color scheme and code.

Table 7-2 Hydrant color coding. The NFPA recommends color coding fire hydrants to indicate the expected flow from the main. From the hydrants available, firefighters can pick the one with the most appropriate flow, using a traffic light scheme for ease of memory.

Class	Flow	Color
AA	1500 gpm (6000 L/min) or more	Light blue
A	1000–1499 gpm (4000–5999 L/min)	Green
B	500–999 gpm (2000–3999 L/min)	Orange
C	Less than 500 gpm (2000 L/min)	Red

Example: What happens if you attempt to conduct a drafting test without consulting the requirements outlined in Table 7-1? Suppose you conduct a draft test on a 2,000-gallon (7,570-L) pump, but instead of using the required two 6-inch (150-mm) intakes, you use only one 6-inch (150-mm) hose.

The calculation for fluid velocity in a line is as follows:

$$V = \frac{\text{gpm} \times 0.3208}{\text{The area of the inside of the hose}}$$

Let's take a pumper with a 1,250 gpm pump. It should have an intake hose not less than 6 inches in diameter. A 6-inch hose has an inside area of 28.27 square inches ($A = \pi \times r^2$). Substituting this area into the velocity formula:

$$V = \frac{1250 \times 0.3208}{28.27} = 14.18 \text{ feet per second}$$

Any velocity faster than 18 feet per second risks cavitation. This is because the only way to make a fluid flow faster is to reduce the pressure on that fluid, but if the pressure on the water were reduced too much, it would boil inside the hard suction hose. Water will boil at a higher than normal temperature if a pressure is placed on the water. For example, a radiator cap increases the pressure within the radiator so that the water may reach a very high temperature without boiling. Conversely, if the pressure is reduced, the water will boil at a lower than normal temperature. This can be demonstrated by creating a vacuum of only 25 inches of mercury. At this reduced pressure, water will boil at 98.7° F (37° C) — just above body temperature. When water in the intake line boils, it creates air bubbles that break against the sides of the intake line, causing pits in the line (cavitation).

Now let's examine the term *net pump pressure*. This is the amount of pressure caused by the pump flow.

Example: A pump is connected to a water hydrant, and at a certain flow, the pressure coming into the inlet of the pump is 20 psi (kPa). This is referred to as the residual pressure. The pump outlet gauge senses a pressure of 170 psi (kPa). Therefore the net pump pressure would be 150 psi (1,035 kPa).

Net pump pressure = outlet pressure – residual pressure

If you are at draft and the gauge on the pump panel reads 150 psi (1,035 kPa), that is not the technically correct reading, for you must take into account the pressure that was lost in getting the water to the pump inlet (the amount of energy that the engine and fire pump used to get the water up to the fire pump).

Using Table 7-3, look up a 1,250-gpm (4,732-L/min) pump with a 6-inch (150-mm) suction hose and a lift from the water level to the center of the pump impeller eye of 10 feet (3 m). From the chart you can see that a 1,250-gpm (3.79-L/min) pump with one 6-inch (150 mm) inlet will have a friction and entrance loss (FEL) of 5.2 feet (1.6 m). Now add the FEL to the distance from the water to the pump eye (this is called static lift), in this case 10 feet. Now 5.2 feet (1.6 m) plus 10 feet (3 m) equals 15.2 feet (4.6 m). This 15.2 feet (4.6 m) is the dynamic lift.

FEL + static lift = dynamic lift

Substituting:

5.2 feet (1.6 m) + 10 feet (3 m) = 15.2 feet (4.6 m)

Remember that 1 psi (7 kPa) equals 2.035 inches (5.17 cm) of mercury (mercury can be used to measure either a pressure or a vacuum). Dividing the 15.2 feet (4.6 m) of dynamic lift by the 2.035 inches (5.17 cm) of mercury (Hg) gives a vacuum reading of 7.47 psi (51.5 kPa) (vacuum). If the pump panel outlet pressure gauge reads 150 psi (1,035 kPa), you must add

Table 7-3 Friction and Entrance Loss in 20 ft (6 m) of Suction Hose, Including Strainer

Flow Rate (gpm)	Suction Hose Size (inside diameter)									
	3 in.		3½ in.		4 in.		4½ in.		5 in.	
	ft water	in. Hg	ft water	in. Hg	ft water	in. Hg	ft water	in. Hg	ft water	in. Hg
250	5.2 (1.2)	4.6								
175	2.6 (0.6)	2.3								
125	1.4 (0.3)	1.2								
300	7.5 (1.7)	6.6	3.5 (0.8)	3.1						
210	3.8 (0.8)	3.4	1.8 (0.4)	1.6						
150	1.9 (0.4)	1.7	0.9 (0.2)	0.8						
350			4.8 (1.1)	4.2	2.5 (0.7)	2.1				
245			2.4 (0.5)	2.1	1.2 (0.3)	1.1				
175			1.2 (0.3)	1.1	0.7 (0.1)	0.6				
450					4.1 (1.0)	3.6	2.7 (0.4)	2.6		
315					2.0 (0.5)	1.8	1.2 (0.2)	1.1		
225					1.0 (0.2)	0.9	0.6 (0.1)	0.5		
500					5.0 (1.3)	4.4	3.6 (0.8)	3.2		
350					2.5 (0.7)	2.1	1.8 (0.4)	1.6		
250					1.3 (0.4)	1.1	0.9 (0.3)	0.8		
600					7.2 (1.8)	6.4	5.3 (1.0)	4.7	3.1 (0.6)	2.7
420					3.5 (1.0)	3.1	2.5 (0.5)	2.2	1.6 (0.3)	1.4
300					1.8 (0.4)	1.6	1.3 (0.2)	1.0	0.6 (0.1)	0.5
700					9.7 (2.7)	8.6	7.3 (1.3)	6.4	4.3 (0.8)	3.8
490					4.9 (1.1)	4.3	3.5 (0.7)	3.1	2.0 (0.4)	1.8
350					2.5 (0.7)	2.2	1.6 (0.3)	1.4	0.9 (0.2)	0.8

Note: Figures in parentheses indicate increment to be added or subtracted for each 10 ft (3 m) of hose greater than or less than 20 ft (6 m).

Table 7-3 continued.

Flow Rate (gpm)	Suction Hose Size (inside diameter)									
	4 in.		4½ in.		5 in.		6 in.		Two 4½ in.	
	ft water	in. Hg	ft water	in. Hg	ft water	in. Hg	ft water	in. Hg	ft water	in. Hg
750	11.4 (2.9)	9.8	8.0 (1.6)	7.1	4.7 (0.9)	4.2	1.9 (0.4)	1.7		
525	5.5 (1.5)	4.9	3.9 (0.8)	3.4	2.3 (0.5)	2.0	0.9 (0.2)	0.8		
375	2.8 (0.7)	2.5	2.0 (0.4)	1.8	1.2 (0.2)	1.1	0.5 (0.1)	0.5		
1000			14.5 (2.8)	12.5	8.4 (1.6)	7.4	3.4 (0.6)	3.0		
700			7.0 (1.4)	6.2	4.1 (0.8)	3.7	1.7 (0.3)	1.5		
500			3.6 (0.8)	3.2	2.1 (0.4)	1.9	0.9 (0.2)	0.8		
1250					13.0 (2.4)	11.5	5.2 (0.9)	4.7	5.5 (1.2)	4.9
875					6.5 (1.2)	5.7	2.6 (0.5)	2.3	2.8 (0.7)	2.5
625					3.3 (0.7)	2.9	1.3 (0.3)	1.1	1.4 (0.3)	1.2
1500							7.6 (1.4)	6.7	8.0 (1.6)	7.1
1050							3.7 (0.7)	3.3	3.9 (0.8)	3.4
750							1.9 (0.4)	1.7	2.0 (0.4)	1.8
1750							10.4 (1.8)	9.3	11.0 (2.2)	9.7
1225							5.0 (0.9)	4.6	5.3 (1.1)	4.7
875							2.6 (0.5)	2.3	2.8 (0.6)	2.5
2000									14.5 (2.8)	12.5
1400									7.0 (1.4)	6.2
1000									3.6 (0.8)	3.2
2250										
1575										
1125										

Note: Figures in parentheses indicate increment to be added or subtracted for each 10 ft (3 m) of hose greater than or less than 20 ft (6 m).

Table 7-3 continued.

Flow Rate (gpm)	Two 5 in.		Two 6 in.		Three 6 in.		8 in.		Two 8 in.	
	ft water	in. Hg	ft water	in. Hg	ft water	in. Hg	ft water	in. Hg	ft water	in. Hg
1500	4.7 (0.9)	4.2	1.9 (0.4)	1.7						
1050	2.3 (0.5)	2.0	0.9 (0.3)	0.8						
750	1.2 (0.2)	1.1	0.5 (0.1)	0.5						
1750	6.5 (1.2)	5.7	2.6 (0.5)	2.3						
1225	3.1 (0.7)	2.7	1.2 (0.3)	1.1						
875	1.6 (0.3)	1.4	0.7 (0.2)	0.6						
2000	8.4 (1.6)	7.4	3.4 (0.6)	3.0			4.3 (1.1)	3.8		
1400	4.1 (0.8)	3.7	1.7 (0.3)	1.5			2.0 (0.6)	1.8		
1000	2.1 (0.4)	1.9	0.9 (0.2)	0.8			1.0 (0.3)	0.9		
2250	10.8 (2.2)	9.5	4.3 (0.8)	3.8	2.0 (0.5)	1.8	5.6 (1.4)	5.0	1.2 (0.4)	1.1
1575	5.3 (1.1)	4.7	2.2 (0.4)	1.9	1.0 (0.2)	0.9	2.5 (0.9)	2.2	0.6 (0.2)	0.5
1125	2.8 (0.5)	2.5	1.1 (0.2)	1.0	0.5 (0.1)	0.5	1.2 (0.4)	1.1	0.3 (0.1)	0.3
2500	13.0 (2.4)	11.5	5.2 (0.9)	4.7	2.3 (0.6)	2.0	7.0 (1.7)	6.2	1.5 (0.4)	1.3
1750	6.5 (1.2)	5.7	2.6 (0.5)	2.3	1.2 (0.2)	1.1	3.2 (1.0)	2.8	0.8 (0.2)	0.7
1250	3.3 (0.7)	2.9	1.3 (0.3)	1.1	0.6 (0.1)	0.5	1.5 (0.4)	1.3	0.4 (0.1)	0.4
3000			7.6 (1.4)	6.9	3.4 (0.6)	3.0	10.1 (3.0)	9.0	2.3 (0.6)	2.1
2100			3.7 (0.7)	3.4	1.7 (0.3)	1.5	4.7 (1.3)	4.2	1.0 (0.3)	0.9
1500			1.9 (0.4)	1.7	0.9 (0.2)	0.8	2.3 (0.7)	2.1	0.6 (0.2)	0.5

Suction Hose Size (inside diameter)

Note: Figures in parentheses indicate increment to be added or subtracted for each 10 ft (3 m) of hose greater than or less than 20 ft (6 m).

Reprinted with permission from NFPA 1901, *Automotive Fire Apparatus*, Copyright © 1999. National Fire Protection Association, Quincy, MA 02269. This reprinted material is not the complete and official position of the National Fire Protection Association, on the referenced subject, which is represented only by the standard in its entirety.

the 7.47 psi (51.5 kPa) to the pump outlet to get a true reading of 157.47 psi (1,085.7 kPa). In other words, conducting the 150 psi (1,035 kPa) test does not require going to 150 psi (1,035 kPa) at draft. In this example a pump panel gauge reading of only 142.5 psi (982.5 kPa) would pass the test if water flow were correct at the rated engine rpm.

This may seem like a lot of work to calculate such a small difference in pressure. However, if you combine this difference with an engine that needs a tune-up on a day of low air pressure, high relative humidity, and high air temperature, your pump could fail the test and mistakenly be deemed deficient and taken out of service.

Estimating Water Flow

It is possible to estimate the volume of water that can be pumped from a hydrant by observing the relationship between the static pressure readings and the residual pressure readings on the pump intake gauge.

Residual pressure readings are controlled by the friction loss in the water supply piping and hose between the pressure gauge location and the pressure source. When the water in the system is static, the factors that cause friction loss have no effect on the pressure reading. For this reason, the static pressure in your kitchen could be the same as that at the hydrant in your front yard even though the flow potentials at those two locations are very different.

The factors that affect friction loss in a water supply do not usually change during a pumping operation; therefore, it is possible to predict flow potentials based on the first information obtained from the pump intake pressure gauge. (The reason for a change in these factors could be that a reserve pump has been brought on line at a water pumping station to meet the new demand caused by the fire. Before conducting any hydrant test inform the local water works department.)

Tests have shown that a pressure drop of 7 percent or less from static pressure to residual pressure indicates that the water supply system and connection will deliver a total of four times the amount presently flowing without requiring any modification to the total supply system, including the hydrant-to-pump connection. If the residual pressure is 20 percent lower than the static pressure, the flow rate can be tripled. If the pressure drop is 30 percent, the flow can be doubled. If it is 40 percent, any increase in flow must be less than the present flow. For example, if the static pressure is 58 psi (400 kPa) and the residual pressure is 55 psi (380 kPa) at a flow of 264 gpm (1,000 L/min), the system can flow 1,057 gpm (4,000 L/min). If the residual pressure is only 40 psi (280 kPa) at a flow of 264 gpm (1,000 L/min), the maximum flow available will be about 528 gpm (2,000 L/min). A change of hydrant to pump connection hose size will change the maximum flow from the hydrant, but the relationship between pressure drop and flow increase potential will remain the same.

Electronic Fire Pump Controls

The older, mechanically controlled diesel engines could not compensate for changes in altitude, barometric pressure, or air temperature. They ran very smoky when traveling over a mountain pass or pumping. (Appendix C on the companion CD-ROM describes the operation of pump controls for nonelectronic engines. This information is also available on the Fire Research Corporation's web site at www.fireresearch.com.) On the newer, electronically controlled engines, sensors inform the computer of the incoming air temperature, the barometric pressure, the temperature of the diesel fuel, and the temperature of the water in the engine block. These all will affect the way the engine runs, but they will not increase its performance at higher altitudes. In other words, a modern electronically controlled engine that has 300 horsepower at sea level will have less than 300 horsepower at 5,000 feet (1,524 m) above sea level; it just will not smoke as much as the older engines did at that elevation. If the apparatus must perform at locations above 2,000 feet (610 m), you must inform the manufacturer during the bidding process.

Electronic pump controls use a set of input signals to control a set of output signals (Figure 7-2). A typical input device is the paddle wheel flow sensor (Figure 7-3). Anytime a magnet rotates past a piece of wire, a voltage is induced into that wire. This induced voltage is an alternating current (AC). The faster the paddle wheel turns, the higher the frequency of the voltage.

Figure 7-2 Combination flow and pressure gauge. The pressure transducer must be downstream from the valve, and the flow sensor must be upstream from the valve. It is important that the pump operator know the pressure in the fire hose, which could be lower than the pressure upstream of the valve.

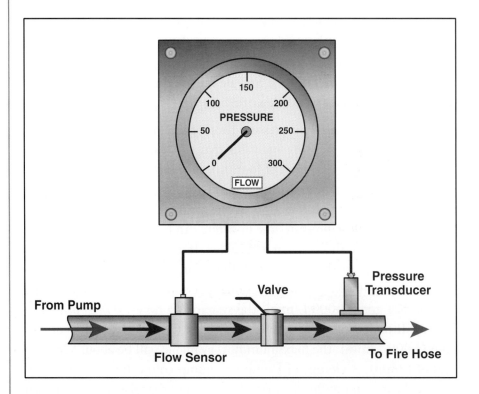

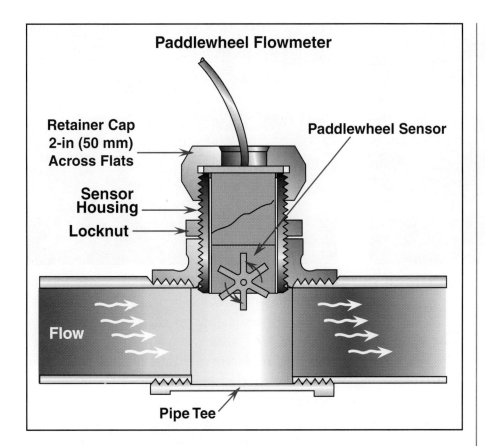

Figure 7-3 Paddlewheel sensor. (Courtesy of Fire Research Corporation)

Paddlewheel Flow Sensor Location

Figure 7-4 illustrates recommended placements for sensors in various pipe configurations. Figures 7-5 and 7-6 illustrate saddle-type paddlewheel sensors. They are shown in blue. These sensors are easier to install on retrofits than are the pipe configurations shown in Figure 7-4. The paddlewheel sensor can be inserted through an appropriate sized hole cut in the red pipe and then secured with the saddle-clamp.

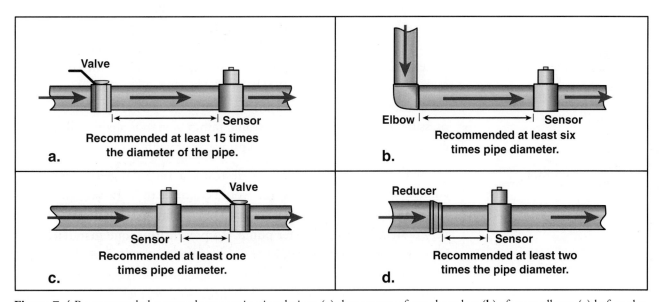

Figure 7-4 Recommended sensor placement in pipe design: (**a**) downstream from the valve; (**b**) after an elbow; (**c**) before the valve; (**d**) after a reducer.

Figure 7-5 Saddle-type flow sensor (in blue). This type is very easy to retro-fit into older units.

Figure 7-6 Three saddle-type flow sensors, one on each pump outlet.

The Role of the CPU

The flow sensors or pressure transducers send their input signals to the central processing unit (CPU) in the electronic pump governor's computer, and then the CPU sends output signals to various devices according to a programmed set of parameters. For example, if the operator were to set the controls for a water pressure of 150 psi (an input), the CPU might direct the engine to increase its rpm to achieve this water pressure (output). If this pumper used an electronic engine, then the electronic pump governor's CPU would achieve the rpm increase by instructing the engine's CPU to lengthen injector "on" time and, therefore, the amount of fuel going to the engine's combustion cylinders. After the engine rpm increased, the electronic pump governor's CPU would look for a corresponding increase in pump pressure. This input signal would come from a transducer. (A transducer is any device that converts an existing form of energy into an electronic signal. In this case, the transducer would convert water pressure to voltage.) The electronic pump governor's CPU can correlate this voltage increase to a corresponding pressure (see Appendix D on the companion CD-ROM; this information is also available on the Class 1 web site at http://www.class1.com/manuals.asp). As more hose lines are opened, the water pressure will drop and the electronic pump governor again will increase engine speed to maintain the 150 psi water pressure set by the operator.

If the water supply were to decrease, a very dangerous condition could develop. The engine speed would continue to increase even though no more water supply was available. This condition, called cavitation, is very destructive to the water pump. To prevent this, one manufacturer's pressure control system will instruct the engine to return to idle speed if during pumping the water pressure measured by the transducer reads less than 25 psi longer than 5 seconds. Check your manual for your system's individual specifications.

The voltages and currents used in both inputs and outputs are very small, in most cases in the millivolt and milliamp range. It is very important that electrical connections be clean and tight to ensure a good electrical connection that will have a low voltage drop. High voltage drops caused by corroded connections will cause false signals and readings. Remember that it does not matter whether the voltage drop occurs on the positive side or on the negative side; it is still a voltage drop, so you also must ensure that negative ground connections are clean. To ensure reliable operation, connect the electronic pump governor to the same 12-volt power source as the engine's CPU. In my experience, most electrical problems occur in the connections. The electronic devices cause very little trouble if they make it past the first 50 hours of use. This is often called the "burn-in time" for electronic devices. Failures during this time are very rare; failures after this time are even rarer.

Electronic Pump Controls for Nonelectronic Engines

Nonelectronic engines, sometimes called mechanical engines, are not as readily available as they once were. The reason is that electronic engines are more environmentally friendly. They are easier to control and produce less pollution than nonelectronic engines. It is possible, however, to have an electronic fire-pump governor on a nonelectronic engine. Many of the components are the same as for the electronic engine. The main differences are in the way the governor receives information from the engine and in the way the governor increases engine speed (output). One manufacturer uses the R terminal on the engine's alternator as the input signal for engine rpm. This requires an initial calibration to set the idle and full speed rpm and a recalibration anytime an alternator pulley combination is changed. In most cases the R terminal is not used for normal alternator operation and is therefore available for this use. The electronic pump governor can read an increase or decrease in engine speed from this terminal (see Appendix C on the companion CD-ROM for a description of pump controls for nonelectronic engines; this information is also available on the Fire Research Corporation's web site at www.fireresearch.com).

A throttle servomotor increases or decreases the engine speed to match the correct fire pump speed for the required water pressure (Figure 7-7). This servomotor pulls or pushes a rod or cable that changes the position of the engine's fuel control rack. The corresponding change in the amount of fuel changes the engine speed. This system still uses a water pressure transducer to input water pressure to the governor from the outlet water manifold.

Interlock An interlock must be provided. In this case, the interlock prevents the governor from operating unless the pump is engaged. If the engine speed is increased from idle and the transmission is in a forward gear, the brakes most likely will be unable to hold the vehicle. This results in a very dangerous situation.

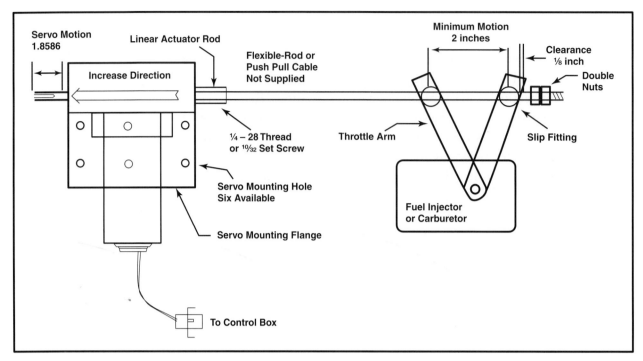

Figure 7-7 Servomotor.

Engine Speed Control An engine speed control can be combined easily with an electronic pump governor. The engine speed control has some excellent uses for the fire service. When not in pump mode but at the scene of a fire, the operator can increase the engine speed to boost alternator output to power the necessary lights without depleting the batteries' reserves. In cold-weather climates this could keep the engine warm, or in warm-weather climates it could keep the air-conditioner fully functioning. It is possible to preset the needed engine speed, and this information could be part of your department's standard operating procedures (SOP). The SOP might read that any time the operator is not in the truck, the engine's idle shall be turned up to 1200 rpm instead of the normal engine idle speed of 600 to 700 rpm. This higher idle speed also would improve the functioning of the engine's cooling and lubrication systems.

Test Categories

The two general categories of tests performed on fire service apparatus are preservice tests and service tests. Preservice tests, sometimes called pre-delivery tests, are done before the apparatus is put into active service at a department. Service tests are performed after the unit is delivered (Figure 7-8). Chapter 14 of NFPA 1901 outlines the standards that must be met for these tests.

Preservice Tests

Preservice tests include the manufacturer's test, hydrostatic tests, certification tests, and acceptance tests. All of these tests must be passed at any

Figure 7-8 (a) Pump test stand used by Edmonton Emergency Services Department. **(b)** The same pump test stand, showing where the water used during testing is recycled into a large underground tank.

elevation up to 2,000 feet (610 m) above sea level, with ambient air temperature between 32° F and 110° F (0° C and 43° C), water temperature between 35° F and 90° F (2° C and 32° C), and on grades up to but not exceeding 6 percent. If the apparatus is intended to perform outside these ranges, the department must inform the manufacturer before the truck is built (Figure 7-9). These tests will apply to a fire pump of at least 750 gpm (2,850 L/min) or greater.

Service Tests

A service test should be performed (a) whenever any system of the vehicle has undergone a service or repair that might affect the ability of the apparatus to pump, (b) whenever the operator even suspects a problem, or (c) at least annually. Most likely the replacement of a wheel bearing, a steering adjustment, or a change of engine oil or batteries would not be cause for a service test. On the other hand, the replacement of an engine water pump, a tune-up of the valves, or a replacement of a fuel injector would require a service test. Figure 7-10 provides a worksheet to use for pump service testing.

Test Requirements

Hydrostatic Tests

Manufacturers must perform a preservice hydrostatic test. This test is performed after the pump has been assembled onto the truck frame. The pump and connecting piping shall be tested to a pressure of 250 psi (1,725 kPa) (NFPA 1901, 14-13.8). The following valves must be closed: the tank-to-pump fill line, the bypass line (if so equipped), and the tank-fill line from the pump. All discharge valves must be open and the outlets

Figure 7-9 Environmental conditions for pump testing may not always be favorable, but they should reflect the conditions in which the pumps may have to operate. (Courtesy of Andy Holzli)

Pump Service Test Worksheet

Test information

Owner _____ Location of test _____ Altitude above sea level _____

Unit number _____ Engine HP/KW _____ @ RPM _____

Date _____ Air temperature _____ Barometric air pressure _____

Pump size _____ Pump type (single or dual stage) _____

Make of pump _____ gear ratio engine to pump _____

Suction hose size used for draft _____

Number of lengths of suction hose _____ Lift from draft pit _____

Test done from Draft ____ Hydrant _____

Water temperature _____

Test results

1. Maximum engine no load RPM _____ Pass/Fail (circle one)

2. Dry vacuum test

    Maximum vacuum attained _____ after _____seconds Pass/Fail

    After 5 minutes _____ Pass /Fail

3. Priming test in _____ seconds Pass/ Fail

4. Water tank to pump flow test, time in seconds _____ Pass/Fail

5. Pressure governor or relief valve test

    @ 150 psi (1034 kPa) _____ Pass/Fail

    @ 90 psi (620 kPa) _____ Pass/Fail

    50% capacity @ 250 psi (1724 kPa) _____Pass/Fail

6. Capacity and overload test:

	Flow	Rated Pressure	Credit Pressure	Net Pressure	ULC/UL RPM	Test RPM	TestFlow/ Pressure	Pass/ Fail
100%								
70%								
50%								
Overload								

7. Engine restart test Pass/Fail

    Run engine at idle for 5 minutes after last test, shut off engine and try to restart. This tests engine starter for heat soak.

8. Maximum engine oil temperature on gauge during test _____

9. Maximum engine coolant temperature on gauge during test _____

Figure 7-10 Worksheet for pump service testing.

capped. All nonvalved intakes must also be capped, and all intake valves that can be closed must be closed. The test shall last three minutes. One major manufacturer of fire pumps hydrostatically tests its fire pumps, valves, and any piping on the pumps to a hydrostatic pressure of 600 psi (4,137 kPa) before they leave the factory. This is to ensure that only quality units are shipped.

Certification Tests

Certification tests are conducted at the manufacturers' facilities and are certified by the contractor. The fire department may order a third party either to conduct or supervise these tests. A number of independent organizations such as Underwriters Laboratories (UL in America, ULC in Canada) conduct these tests. It is very important that the certification tests are done by an independent organization in order to protect both the buyer and the manufacturer. Even more important than integrity, the independent certifier will bring to the test site properly sized and calibrated test gauges (Figures 7-11 and 7-12). NFPA 1901, 14-13.2.2.4, provides more information on the standards for these gauges (for more information on electronic controls, see Appendices D–J on the companion CD-ROM; this information is also available on the Class 1 web site at http://www.class1.com/manuals.asp).

The certification tests should include at least the pumping test, the pumping engine overload test, the pressure control system test, and the vacuum test. When the tests have been completed, the manufacturer must attach to the operator's panel a test plate that states the rated discharge and pressures together with the speed of the engine determined during the certification tests.

Figure 7-11 A pitot tube is used to measure the velocity pressure of a stream of water. This velocity pressure is used to calculate the amount of water flowing from a discharge outlet.

Figure 7-12 Clamping the pitot tube directly to the test nozzle is safer and provides more accurate readings than holding it.

Test Conditions The conditions for certification tests are very wide ranging. Air temperature may be between 0° F and 101° F (–18° C and 43° C); water temperature may be between 35° F and 90° F (2° C and 32° C); and barometric pressure must be at least 29 inches of Hg (98.2 kPa) (this can be corrected to sea level). All engine-driven accessories must be connected and operational during the tests. You cannot cheat and disconnect the alternator. The alternator must be providing the total connected load (TCL) during these tests (see Chapter 6 for an explanation of TCL).

One of the greatest problems with fire truck design is heat generation/dissipation; therefore, you cannot remove any panels, grills, or gratings that are not meant to be opened during pumping to alleviate this problem. Nor can you lift the hood if the engine overheats during the tests. During these tests, the engine, transmission, and any other parts cannot exhibit undue heating, loss of power, or any other defect. With the large variation of air and water temperature allowed in the standard, a truck could very easily pass the tests at an air temperature of 32° F (0° C) and a water temperature of only 40° F (4.4° C), but these readings may not represent the conditions that the truck might experience on a normal summer's day in your location. It is the department's responsibility to inform the manufacturer of the conditions that must be in place during the acceptance tests. A reputable manufacturer may be able to install an engine fan with more blades, a radiator with more cores, or make other changes to address the special climactic conditions specified for the tests (Figure 7-13). Be up front with manufacturers during the bidding process so that they will have time to alter the design to ensure the apparatus will pass the preservice tests under the specified conditions.

The electronic tachometer operated from the engine or transmission can be used to measure engine rpm. If you are unsure of this reading's accuracy, connect a mechanical drive tachometer to the port provided and use that reading (Figures 7-14 and 7-15). The mechanical tachometer installed on the pump panel or in the cab is not considered accurate enough for these tests. Accuracy must be within ±50 rpm at the actual speed being measured.

> **Note:** The accuracy of a photo tachometer is easy to demonstrate. A standard electrical grid system for most North American cities operates at 60 Hz (also called 60 cycles per second). A fluorescent light bulb, therefore, turns on 60 times a second and turns off 60 times a second, or 3,600 times a minute. Thus, the photo tachometer will read 7,200 rpm when directed at a fluorescent light.

Pumping Test The first series of tests will take at least three hours, consisting of a continuous 2-hour test followed by two 30-minute tests. During the continuous 2-hour test, the pump must operate at its rated capacity

Figure 7-13 Reversible fan blades on a front-mount pumper can be switched in winter to keep the pump and the operator warm. (Courtesy of Hill's Hot Shot Service, Ltd., Sherwood Park, Alberta)

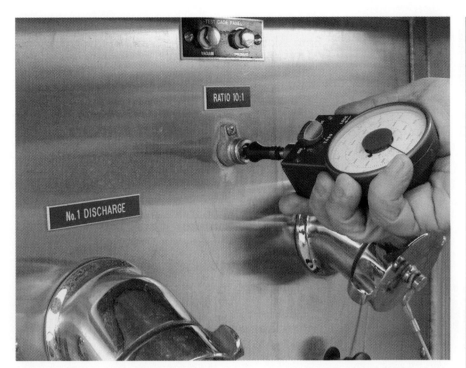

Figure 7-14 Mechanical drive tachometer connected to the pump through the appropriate port. Note the 10:1 ratio.

Figure 7-15 Exceeding the manufacturer's recommended maximum rpm can have devastating results. Note the hole in the engine block caused by connecting rod failure.

(flow) at a net pump pressure of 150 psi (1,035 kPa). During this time you cannot stop the unit to add fuel. The next test lasts 30 minutes and requires the pump to operate at 70 percent of rated capacity (flow) and a higher pressure of 200 psi (1,380 kPa). The last test requires the pump to operate 30 minutes at 50 percent of rated capacity (flow) and a pressure of 250 psi (1,725 kPa) (Figure 7-16). During the 70-percent and 50-percent tests, you may need to switch the pump from parallel mode to series mode

Figure 7-16 Front-drive failure between the engine and the pump gear box on a front-mount pump. This type of failure can occur during the 50 percent, 250 psi (1,725 kPa) test. (Courtesy of Hill's Hot Shot Service, Ltd., Sherwood Park, Alberta)

if it is a two-stage pump. During the last two tests adding fuel is permissible, but you may want to ask the manufacturer how long the truck's fuel tank capacity will allow it to run, so that you will have an idea of how long your truck really can pump.

> **Note:** Talk to the manufacturer if you think you require more fuel capacity. However, installing a dual tank system is not advisable.

Overload Test If the pump's capacity is 750 gpm (2,850 L/min) or larger, an overload test must be performed immediately after the 150-psi (1,035 kPa) test. The pump is run for 10 minutes at a pressure of 165 psi (1,138 kPa). During all of the above tests the discharge pressure, intake pressure, and most importantly, engine speed (rpm) must be recorded.

If the pump's capacity is less than 750 gpm (2,850 L/min), the overload test is not required, and the other tests' times are reduced. The entire test shall now be 50 minutes, with the 100-percent capacity test at 150 psi (1,035 kPa) lasting 30 minutes, the 70-percent test at 200 psi (1,380 kPa) lasting 10 minutes, and the 50-percent test at 250 psi (1,725 kPa) also lasting 10 minutes.

Pressure Control System Test The pressure control system test will be done if the fire pump has either a very simple relief valve or a very complex electronic engine governor (for more detailed information on the function and maintenance of relief valves see Appendix K).

The pump must be operated from draft at three different pressures in this order: 150 psi (1,035 kPa) ±5 percent, 90 psi (620 kPa) ±5 percent, and 250 psi (1,725 kPa) ±5 percent. This test is to ensure that the systems' water pressure does not rise (spike) by more than 30 psi (207 kPa) as discharge valves are closed. Discharge valves cannot be closed faster than within 3 seconds or slower than within 10 seconds. If the water pressure were to spike as discharge valves were closed, it might be very difficult for firefighters to maintain their grip on the hose lines. A loose charged hose line is extremely dangerous. Therefore, this test has very important safety consequences.

Vacuum Test Priming devices should be tested as follows. With all intake valves open and capped (or plugged) and all discharge valves closed and discharge caps removed, operate the priming pump to create a vacuum of 22 inches of Hg (74.5 kPa) in the pump body and then stop the priming pump. The vacuum must not drop by more than 10 inches of Hg (33.9 kPa) in 5 minutes. The 22 inches of Hg (74.5 kPa) requirement applies to tests done at elevations from sea level to 2,000 feet (610 m) above sea level. For every 1,000 feet (305 m) above 2,000 feet (610 m), the vacuum standard is reduced by 1 inch of Hg (3.4 kPa); for example, at 5000 feet

Pump Packing Adjustment

Fire pumps use two types of seals for the pump shaft: mechanical seals or stuffing box seals. Mechanical seals require no adjustment during their service life (**Figure 7-17**). If leakage is discovered, they must be replaced. Mechanical seals give very long life and are almost trouble free.

More common are the stuffing box seals (also called split-type packing glands) (**Figure 7-18**). They require frequent adjustment over the life of the pump. Some manufacturers use two stuffing boxes, one on each end of the shaft. Some use only one stuffing box on the drive end.

If the packing gland is too tight, the pump shaft will overheat and score as the water is used to cool the shaft. If the packing is too loose, then the pump will not be able to draft, as the primer will not be able to create a low enough vacuum to prime the pump. It is a common misconception that the water that drips out of the shaft packing comes from the small amount of water that leaks past the shaft from the pump impeller. In fact, this is not the case.

External plumbing routes high-pressure water in the amount of gallons or liters per minute from the pump's water outlet to the shaft and back to the pump's inlet (**Figure 7-19**). Lantern rings form a space for this water to flush by the shaft and cool it. If you were simply to stuff more packing into the stuffing box, the lantern rings could be pushed out of their proper location, cutting off the water flow. Manufacturers all have their own methods to determine when the packing needs to be adjusted and their own procedures for making the adjustments; however, some general guidelines apply. Operate the pump from draft, hydrant, or water tank; circulate the water; and do not allow the water to overheat. With between 100 psi (689 kPa) and 150 psi (1034 kPa) in the pump system, count the number of water drops coming out of the stuffing boxes. As few as 8 drops per minute and as many as 120 are common; check the service manual for your pump.

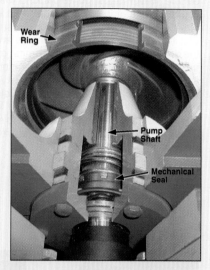

Figure 7-17 Cutaway of a mechanical seal.

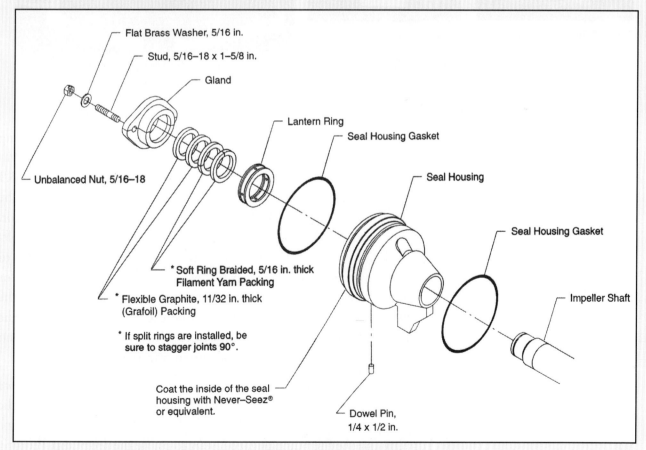

Figure 7-18 Stuffing box seal. Note the lantern ring. (Courtesy of Waterous Company)

If stuffing-box adjustment is necessary, shut off the pump and engine. Never adjust pump packing with the engine running. You could be caught in the driveshaft and killed or seriously injured. Make the adjustment in small steps, usually one adjustment notch at a time, and then start the engine and fire pump and recheck the drip rate. This is a time-consuming procedure; do not rush and possibly overtighten the packing. The one exception to the above rule is a Darley fire pump with injection type packing (**Figures 7-20, 7-21, and 7-22**). This packing can be adjusted from outside the fire truck body with the proper accessory kit. If your fire pumps use this type of packing, you will be able to make the adjustment from a safe distance with the engine and fire pump operating.

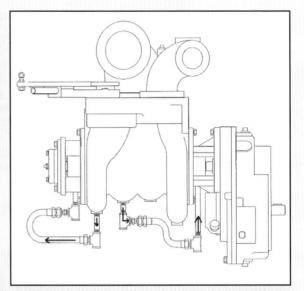

Figure 7-19 Water flow for seal cooling. (Courtesy of Waterous Company)

Figure 7-20 Cutaway model of a Darley injection packing with external adjusting nut.

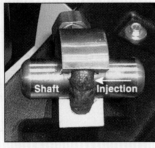

Figure 7-21 Cutaway model of a Darley injection packing seal. **Note:** Hardened shaft where packing seal is seated.

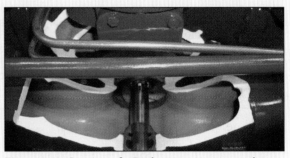

Figure 7-22 Cutaway of a Darley injection type seal packing.

(1,525 m) you need to develop only 19 inches of Hg (64.3 kPa). During the test, the priming pump cannot be used. Any excess vacuum leaks will cause the vacuum to drop faster than the allowed 10 inches of Hg (33.9 kPa) in 5 minutes. If the primer is electrical, running the engine may be necessary to provide enough electrical energy to power the primer motor. These motors can draw in excess of 250 amps at 14 volts. If battery voltage were to fall to, say, 11 volts, the priming pump might not be able to pass these vacuum tests. You may not exceed engine-governed speed for this test. If you are using an engine vacuum priming system, running the gasoline engine at full throttle will be of no use because available engine vacuum decreases as engine rpm increases. Once you know there are no excess vacuum leaks, move on to the pump priming test.

Pump Priming Test To perform a pump priming test, connect no more than 20 feet (6 m) of suction hose, with a proper strainer attached, to the fire pump. Suspend the suction hose in a drafting pond. With the engine running and the fire pump engaged, start to prime the fire pump. Priming pumps of 1,250 gpm (4,732 L/min) or less should take no longer than 30 seconds. For pumps of 1,500 gpm (5,678 L/min) or more, the allowable time is 45 seconds. If your system uses an auxiliary intake of 4 inches (100 mm) or larger or an intake pipe having a volume of 1 cubic foot (0.0283 m³) or more, then you can add 15 seconds to the allowed priming time.

A pump is considered to be primed when enough air has been removed and replaced with water that the pump can operate on its own. Some people mistakenly think a pump is primed when water comes out of the priming pump's discharge hose and lands on the ground; this is not true. When you have successfully primed the pump, water will be flowing out of a discharge port.

Water-Tank-to-Pump Flow Test The water lines from the water tank to the fire pump must be large and straight. Because the initial water supply for fighting many fires usually comes from the tank, it is important to understand that the lines from the water tank to the fire pump will be much smaller than the intake lines for drafting. Therefore, the fire pump can never put out its rated capacity using the water tank.

How much water can you expect from a water line? Many feed lines from the water tank to the pump are 3 inches (76 mm) in diameter, but that same fire pump may need a 6-inch (152-mm) diameter intake to successfully draft and meet its capacity rating. Is a 6-inch diameter (152-mm) intake twice as big as a 3-inch (76-mm) intake? To calculate the area of a 3-inch line's opening,

$$A = \pi r^2$$
$$A = \pi (d/2)^2$$

Substituting:

$$A = 3.1416 \times 1.5 \times 1.5$$
$$A = 7.069 \text{ in.}^2$$

The area of the inside of a 6-inch intake is as follows:

$$A = 3.1416 \times 3 \times 3$$
$$A = 28.2744 \text{ in.}^2$$

Therefore, a 6-inch intake line has an opening four times larger than a 3-inch intake line, which means it has the capacity for four times the water flow.

If the water tank is certified to be less than 500 gallons (1,900 L), then the piping and valves must be able to deliver at least 250 gpm (950 L/m) during the test. If the water tank is 500 gallons (1,900 L) or greater, then the required minimum rate is 500 gpm (1,900 L/m) or the rated capacity of the pump, whichever is less.

To prepare for the test, position the truck on level ground (not to exceed a 6 percent grade) and fill the water tank to overflowing. Close all intakes to the fire pump, including the tank fill line and the bypass cooling lines. Connect as many discharge lines and hoses as necessary to match the expected pump discharge rates. Start the engine and engage the fire pump. Now open the tank-to-pump valves and the discharge valves to the hose lines and open the nozzles. Open the engine throttle until rated flow is established (±5 percent), and record the discharge pressure. With the engine running, slowly close the discharge valves and open the bypass lines to keep the water in the pump from overheating and refill the water tank.

Now begin the test. Open the discharge valves again and either note the time or start a stop watch. You may have to change the engine throttle to maintain the discharge pressure. Keep recording the time until the discharge pressure drops 5 psi (34 kPa); this indicates you are running out of water. Idle down the engine and disengage the fire pump. To pass this test, the pump must maintain the rated flow until it has removed no less than 80 percent of the water from the tank. To calculate the volume of water discharged, multiply the rate of discharge by the time elapsed from when you open the discharge until you read the 5-psi (34 kPa) drop in pressure.

Example: For a water tank with a capacity of 1,500 gals (5,678 L), a fire pump with a capacity of 1,000 gpm (3,785 L/min) should be able to deliver 500 gpm (1,900 L/min). Eighty percent of 1,500 gals (5,678 L) is 1,200 gals (4,525 L), and 1,200 (4,525) divided by 500 (1,900) is 2.4. Thus, this fire pump should be able to deliver water at the rated flow of 500 gpm (1,900 L/min) for 2.4 minutes, or 144 seconds.

Solving Pumping Problems

If any operational problems arise during any of the tests described above, the most efficient way to identify the cause is to adopt a systematic and logical approach based on elimination. An experienced pump operator will often be able to recognize the symptoms of the common faults and pinpoint the cause, but if the anticipated difficulty is not the actual problem, the time spent on it is wasted.

The D-S-O Sequence

The logical process when diagnosing an operational problem is to follow the power flow from engine to pump impeller and the water flow from source to nozzle. To remember the sequence, think of the key words or letters: drive, supply, output (D-S-O).

<u>D</u> Drive: is the pump running at the correct speed?

<u>S</u> Supply: can water enter the pump, air free and in sufficient volume?

<u>O</u> Output: does anything hinder the required pressure and volume from getting out of the pump?

> **Note:** You may want to impress upon your firefighters how little water they actually have available when using the water tank.

Drive The term *drive* refers to all things involved in the mechanics of turning the pump. To check the drive system answer the following questions.

If the pumper is a front mount:

- Is the main transmission in neutral?
- Is the pump transmission in pump gear?
- Is the pump actually turning?

If the pumper is a midship mount:

- Is the main transmission in the correct gear?
- Is the pump transmission in pump gear?
- Is the transfer valve in the correct stage (if multistage)?
- Is the pump turning at the right speed? (Check the speedometer.)

If the pump is a PTO drive:

- Is the main transmission in neutral?
- Is the PTO fully engaged?
- Is the pump shaft turning?

Supply The term *supply* refers to all equipment and conditions involved with the delivery of water into the pump. To check the supply, answer the following questions.

If supply is at draft:

- Is the correct suction hose being used?
- Is the strainer located at the correct depth?
- Are there any air leaks in the suction line?
- Are there any air leaks in the truck?
- Is there any blockage of any of the strainers?
- Is the intake valve fully open (if so equipped)?
- Are there any air traps in the suction hose?
- How high is the lift?
- Did the primer operate correctly?
- Are any other valves opened?
- How long is the suction line?
- What is the interior condition of the suction hose?

If supply is a hydrant:

- Is the suction hose large enough for the required flow?
- Is the hydrant fully open?
- Is there a residual pressure?

- Are the intake valves in use fully open?

- Are there any air or water leaks?

- Is the tank valve open?

- Are any intake strainers plugged?

- Are any hose lines kinked?

Output The term *output* refers to all of the equipment involved with the delivery of water to the nozzles at satisfactory volume and pressure. To check output answer the following questions. (Appendix C on the companion CD-ROM describes the operation of pump controls for nonelectronic engines. This information is also available on the Fire Research Corporation's Web site at www.fireresearch.com.)

- Is the governor operating correctly?

- Is the relief valve operating correctly?

- Are the appropriate discharge valves open?

- Can the engine reach the required rpm?

- Is the required flow-pressure within the pump's capacity?

- Are valves open that should be closed?

- Is there any unusual leakage to the ground?

- Are the gauges operating properly?

- Are any hoses kinked?

- Are you attempting to move too much water through the size of valve or hose in use?

Working from the Known to the Unknown

Every time we diagnose a problem in anything we do, we begin a process of elimination. We dismiss anything that we determine is not part of the problem, and continue until we have found the problem.

When a problem's cause is unknown, begin by assembling all of the known information. When you have completed this process, the unknown will become obvious. Often a single piece of information can answer a number of questions, saving a lot of steps. A good example is the speedometer of a midship pumper. If the speedometer is reading in the range from 9 mph to 19 mph (15–30 km/h) at engine idle, you know that all of the correct steps in putting the pump transmission into pump gear have been followed. This eliminates the need to check all of the individual procedures involved in putting the pump into gear. It is important to follow a logical sequence in all troubleshooting; again, use the Drive-Supply-Output sequence in the process of working from the known to find the unknown.

Example: A midship pumper is connected to a hydrant with soft suction, but no water is reaching the nozzle.

1. The first step is to check the *drive* system. The speedometer reads 12.5 mph (20 km/h) at engine idle, so we can assume that all is well with the pump drive unless, of course, there is a mechanical failure in the pump transmission or the pump.

2. The second step is to check the *supply* system. The pump intake pressure gauge reads 65 psi (450 kPa). This tells us that water is getting into the pump (if the gauge is working).

3. Now check the *output* system. The pump-master-pressure gauge reads 145 psi (1,000 kPa). This indicates that the pump is producing pressure within the pump discharge manifold. From the information we have at this point, we know that the problem is somewhere between the pump manifold and the nozzle. The most likely fault in this case would be a closed discharge valve.

The D-S-O system and a good working knowledge of your pumper can make troubleshooting pumper operational problems quick and accurate (Table 7-4).

Summary

NFPA standards for fire-pump testing need to be your bible for conducting tests on your fire apparatus. To interpret the standards and the test results correctly, you must have a thorough understanding of hydraulics and all of the factors that can influence pump performance. Then you must apply that knowledge in a logical sequence to identify problems in pump operations.

Table 7-4 Pump troubleshooting chart.

A. Operating from Draft

Trouble	Problem Cause	Remedy
1. Pump fails to prime	1. Pump not in gear or not in proper gear	1. Check road and pump transmission and select proper gear
	2. Air entering pump through open valves	2. Make sure all bleeder, discharge, intake, and drain valves are closed
	3. Defective gaskets	3. Replace gaskets
	4. Loose couplings	4. Tighten
	5. Clogged intake strainer	5. Clean
	6. Air entering strainer	6. Locate at proper depths
	7. Primer not operating properly	7. Service primer
	8. Primer valve not opening	8. Check linkage movement and adjust as necessary
	9. Pump leakage	9. Adjust or replace packings
	10. Suction lift too high	10. Reduce lift (should not exceed 20 ft [6 m])
	11. Primer float chamber full of water (vacuum primer)	11. Push in primer handle, count to 10, and operate primer again
	12. Engine RPM wrong	12. Set engine to correct RPM for the type of primer being used
2. Pump fails to deliver its rated capacity or pressure	1. Pump not in gear or not in proper gear	1. Check road and pump transmission and select proper gear
	2. Automatic transmission downshifts	2. Activate transmission "lock up" switch
	3. Air entering pump	3. Close all bleeders and drains
	4. Clogged intake strainer	4. Clean
	5. Clogged pump strainer (pump to suction hose)	5. Clean
	6. Vortex	6. Locate pickup strainer at the proper depth
	7. High point in suction line	7. Correct condition
	8. Defective suction hose (loose lining)	8. Replace

Table 7-4 continued

A. Operating from Draft

Trouble	Problem Cause	Remedy
	9. Worn pump wear rings	9. Overhaul pump
	10. Relief valve open	10. Close relief valve
	11. Pump in wrong stage (multi-stage pump)	11. Select proper stage
	12. Engine lacks power	12. Tune up or repair engine
	13. Clogged impeller vanes	13. Backwash pump
	14. Excessive friction loss in suction tube or dry hydrant	14. Get closer to water
	15. Pressure governor keeping engine RPM down	15. a. Check governor b. Drain governor chamber c. Select proper mode d. Clean strainer e. Check for slipping cable f. Check hydraulic fluid level

B. Operating from Hydrant

Trouble	Problem Cause	Remedy
1. Pump will not deliver rated capacity or pressure	1. Pump not in gear or not in proper gear	1. Check road and pump transmission and select proper gear
	2. Automatic transmission downshifts	2. Activate transmission "lock up" switch
	3. Hydrant not fully open	3. Open hydrant
	4. Drain or bleeder valves open	4. Close all drains and bleeders
	5. Clogged pump strainer	5. Clean
	6. Intake or discharge valves	6. Open all valves
	7. Excessive pump packing leakage not fully open	7. Replace or adjust pump packings
	8. Relief valve open	8. Close relief valve
	9. Excessive friction loss between pump and hydrant	9. Make adequate hook up
	10. Pump in wrong stage (multistage pump)	10. Select proper stage
	11. Engine lacks power	11. Tune up or repair engine
	12. Clogged impeller vanes	12. Backwash pump

Table 7-4 continued

B. Operating from Hydrant

Trouble	Problem Cause	Remedy
	13. Pressure governor keeping engine RPM down	13. a. Check governor b. Select proper mode c. Check for slipping cable d. Check for clogged strainer e. Drain governor chamber f. Check hydraulic fluid level

C. Operating from Booster Tank

Trouble	Problem Cause	Remedy
1. No water delivery	1. Pump not in gear or not in proper gear	1. Check road and pump transmission and select proper gear
	2. Tank to pump valve is closed	2. Open tank valve
	3. Air trapped in pump	3. Prime pump
	4. Automatic transmission not in proper gear	4. Activate "lock up" switch
	5. No water in tank	5. Fill
2. Low delivery of water	1. Pump in wrong gear	1. Select proper gear
	2. Automatic transmission downshifting	2. Activate transmission "lock up"
	3. Tank valve not fully open	3. Open
	4. Blockage of supply line	4. Repair
	5. Discharge valve not fully open	5. Open valves
	6. Tank unable to get air	6. Open cover, clear blockage of overflow pipe
	7. Pump leakage	7. Repair as necessary
	8. Relief valve open	8. Close
	9. Clogged impeller vanes	9. Backwash pump
	10. Pressure governor keeping engine RPM down	10. a. Check governor setting b. Drain governor chamber c. Select proper mode d. Clean strainer e. Check for cable slippage f. Check hydraulic fluid level

Table 7-4 continued

D. General Operations

Trouble	Problem Cause	Remedy
1. Gauge not functioning	1. Gauge shut off valve is closed	1. Open valve
	2. Line blocked	2. Service lines and gauges
	3. Gauge unserviceable	3. Replace
2. Low engine RPM	1. Broken or slipping throttle cable	1. Tighten or replace
	2. Pressure governor holding engine RPM down	2. a. Check governor setting b. Drain governor chamber c. Select proper mode d. Clean strainer e. Check for slipping cable f. Check hydraulic fluid level
	3. Engine requires tune up	3. Tune up as necessary
3. Relief valve malfunctions	1. Excessive rise	1. Service unit
	2. Surging	2. Service unit
	3. Excessive throttle required	3. Service unit

Chapter 7 Review Questions

Short Answer

Write your answers to the following questions in the blanks provided.

1. Would warm humid air increase or decrease an engine's ability to produce horsepower? Explain your answer.

   _____

   _____

   _____

2. If normal atmospheric pressure is 14.7 psi (101.2 kPa) at sea level, what should the pressure be at 5,000 feet (1,524 m) above sea level?

   _____

3. How would the decreased air pressure at 5,000 feet (1,524 m) affect an engine?

   _____

   _____

   _____

4. What do pumps make? _____

5. What causes pressure? _____

6. When testing a fire pump that has a rated capacity of 1,250 gallons per minute (4,732 L/min), what is the maximum suction hose size, how many pump inlets are allowed, and what is the maximum lift?

   Suction hose size: _____

   Number of pump inlets allowed: _____

   Maximum lift: _____

7. When referring to a water hydrant, what do the terms *residual pressure* and *static pressure* mean?

   Residual pressure: _____

   _____

   Static pressure: _____

   _____

8. What causes cavitation? _____

9. Describe two effects of cavitation.

   _____

   _____

   _____

10. A hydrant has a residual pressure of 30 psi (207 kPa), and the fire pump has an output pressure of 190 psi (1,310 kPa). What is the net pump pressure?

   _____

11. A 1,000-gpm (3,785-L/min) pump is being tested with one 5-inch (125-mm) suction hose. The lift from the level of the lake to the center of the impeller eye is 10 feet (3 m). What is the FEL, what is the static lift, and what is the dynamic lift?

    FEL: _____

    Static lift: _____

    Dynamic lift: _____

12. Define the term *transducer* and give one example.

    _____

    _____

    _____

13. What advantages do electronic diesel engines have over mechanical diesel engines?

    _____

    _____

14. True or False. An electronic diesel engine produces the same 300 horsepower at 3,000 feet elevation as it does at sea level. If false, explain why.

    _____

    _____

15. What is the role of the CPU on the electronically controlled fire pump?

    _____

    _____

    _____

16. Explain why it is important to have clean connections on all wiring associated with electronically controlled fire pumps.

    _____

    _____

    _____

    _____

17. When is it necessary to recalibrate the engine idle speed and the engine rpm on a nonelectronic engine fitted with an electronic fire-pump governor?

    _____

    _____

18. Why is it necessary to have an interlock on an electronically controlled governor connected to a mechanical engine?

    _____

    _____

    _____

    _____

19. Name the four individual tests that can make up the group of preservice tests.

   _____   _____

   _____   _____

20. When should a service test be performed on a fire pump?

   _____

   _____

21. Manufacturers perform a preservice hydrostatic test with the pump on the truck; what is the minimum pressure at which this test must be performed?

   _____

22. How long does the preservice hydrostatic test last?

   _____

23. Why should an independent certification organization perform the certification tests?

   _____

24. Why must all engine driven accessories remain connected during the certification tests?

   _____

25. How long must a fire pump rated at 1,250 gpm (4,732 L/min) deliver 100 percent of its rated capacity during the pumping test?

   _____

26. At what pressure is the pumping test done?

   _____

27. When, if at all, is it permissible to add engine fuel during this test?

   _____

28. How long must the pump deliver 70 percent of its flow rate at 200 psi (1,380 kPa), and how long must it deliver 50 percent of its capacity at 250 psi (1,725 kPa)?

   70 percent at 200 psi: _____

   50 percent at 250 psi: _____

29. What are the specifications for the overload test?

   Pump capacity: _____

   Pressure: _____

   Duration: _____

   Flow: _____

30. Is the pressure-control test conducted from a hydrant or from draft?

   _____

31. What is the maximum allowable pressure rise during the pressure-control test?

   _____

32. When conducting the vacuum test at sea level, how long do you have to reach the required vacuum?

_____

_____

33. What requirements must a pump meet to pass the vacuum test?

_____

_____

34. How long must the pump be able to hold the required vacuum without dropping more than 10 inches of vacuum (33.9 kPa)?

_____

35. When is a fire pump considered to be primed?

_____

_____

36. Why is the water-tank-to-pump test important?

_____

_____

37. Can a fire pump put out its rated capacity when using the water tank as the water supply? Explain your answer.

_____

_____

_____

38. Which type of pump shaft seal requires no adjustment during its service life?

_____

39. What is another name for a split-type packing gland?

_____

40. What problem is created if too much packing is stuffed into the stuffing box?

_____

_____

41. What is the acceptable range of drops per minute coming from the stuffing box seals?

_____

_____

42. What safety precautions need to be observed when adjusting the stuffing box seals?

_____

_____

43. What is the only exception to the safety precautions named in question 42?

_____

_____

Chapter 8
Transmissions for the Fire Service

Transmissions for the Fire Service

Introduction

The vast majority of transmissions used in the fire service are the automatic type, and the Allison automatic has been the transmission of choice for many years. This transmission is reliable, easy to operate, and very adaptable to the needs of the fire service. Some of its maintenance concerns are different from those of a standard manual gear transmission, and we will concentrate on these. They are common to both the older automatic transmissions with hydraulic controls and the newer automatic transmissions with electro-hydraulic controls.

Automatic Transmissions with Hydraulic Controls

Mechanical automatic hydraulic transmissions use hydraulic pressure to engage the clutch packs. The hydraulic pressure is modified by mechanical inputs such as governor speed, modulator pressure, and mechanical springs and by operator shift lever controls.

Transmission Models

The three models of mechanical transmissions that you are most likely to encounter are the AT, MT, or HT transmissions.

AT 500 Series The AT transmission is used in lightweight units such as ambulances and fast attack units that may be either gasoline or diesel powered. These single-rear-axle four-speed transmissions can weigh as much as 275 pounds (125 kg) without lubricants, and as much as 315 pounds (143 kg) with a PTO or optional retarder. Maximum gross weight for vehicles fitted with these transmissions is 30,000 pounds (13,608 kg).

MT 600 Series The MT transmission is often used in the single-rear-axle fire apparatus unit. This is a very popular transmission that can be used on gasoline or diesel powered units. It comes with either four or five forward ranges. The second digit in the model number indicates each unit's number of forward ranges. For instance, units with four ranges might be designated MT 640 or MT 643, while models with five ranges might be designated MT 650 or MT 653.

A five-speed model can be either a close-ratio (CR) transmission (54CR) or a deep-reduction (DR) transmission. The extra range in the five-speed models is an added planetary gear set installed on the rear of the four-speed transmission. With the correct valve body, this arrangement can give the transmission a deep reduction range (mostly for off-road use) or a close ratio pattern. The closer ratio is most useful with engines that have a narrow torque curve, such as some diesel engines.

The low range in the deep-ratio transmission should not be used as a starting range. To use this range, you should stop the truck, place the transmission in range, and then proceed. When the need for this very low range passes, bring the apparatus to a complete stop, select drive or second range, and then continue automatic upshifts as normal. Do not shift out of or into these low ranges while moving; doing so with this model of transmission will place extreme stress on driveline components.

Distinguishing DR and CR transmissions is easy. Place the transmission in reverse. If you can go faster in reverse than you can in the lowest forward range, you have a DR model. If you can go faster in the lowest forward range than you can in reverse, you have a CR model. A retarder is available on all models of the MT 600.

This transmission uses an internal transmission oil filter. If requested, an external filter can be added.

HT 700 The HT 700 model transmission is used in aerial fire trucks, heavy fire pumpers, and most tandem axle units. These transmissions are very similar in design to the MT 600 series, just larger and suited to vehicles with more engine input power and a heavier gross vehicle weight (GVW). Rather than an internal oil filter, these units use a small screen inside the oil pan and a large external oil filter. They are available as four- or five-speed models, and the option for a CR or DR model is available. The HT can have a retarder, and different PTO (power takeoff) options are available. These options are most useful on aerial fire trucks as they allow for running of hydraulic pumps, foam systems, or high-pressure low-volume water pumps.

When installing these PTOs, use only Allison-approved gaskets. Never use the older style gaskets made of cork; these can leak and may require the use of a gasket sealant. Such gasket sealants are not recommended. If even a small amount of sealant were to get into the transmission fluid, it could cause erratic shifting and valve sticking, with disastrous results.

Figure 8-1 A model MT 643 Allison transmission, serial number 0101818, with an assembly number of 6884350.

Identifying Allison Transmission Models Allison transmissions have an identification plate on their right-hand side. This plate is also called the SAM plate, which stands for <u>s</u>erial, <u>a</u>ssembly, and <u>m</u>odel numbers. The plate shown in Figure 8-1 identifies the transmission as a model MT 643. The assembly number is 6884350, and the serial number is 0101818. There are thousands of different variations of the AT, MT, and HT models. You can use the assembly number together with the parts manual to obtain the following important information:

- Governor number
- Torque converter torque ratio
- Engine rpm for which the transmission was designed
- Whether the transmission has a modulated lockup

Transmission Governor

Each automatic transmission has a governor. The governor provides information to the valve body and is the closest thing that an automatic transmission has to a brain. The governor is driven by the transmission output shaft. The faster the shaft spins, the faster the governor turns. This rotation and the centrifugal force created at the governor produce a governor pressure proportional to the centrifugal force. Governor pressure flows to the shift valves to cause the transmission to upshift or downshift. Different governors will have very different governor pressure profiles and hence must be matched to the correct valve bodies.

Figure 8-2 The governor number of this governor valve is 53.

The modulator valve in the transmission is another information source for the valve body. It obtains its data from the position of the throttle pedal in the cab. The governor pressure and the modulator valve pressure determine when the transmission upshifts or downshifts. A two- or three-digit number is inked on the head of each governor, as shown in Figure 8-2. This number must match the parts book's information for the governor assembly number. If a shifting complaint is received, an incorrect governor may have been installed or the governor may be defective. This will cause the transmission to shift too early (no power) or too late (very harsh shifts). Like any mechanical part, the governor can wear out over time causing erratic shifts. You may wish to keep a replacement governor in stock if you have several transmissions with the same assembly number.

Governor Maintenance Inspect the gear on the governor for wear. The faster the governor turns, the higher its pressure. This pressure tells the shift-signal valve in the valve body when to shift up or down. If the gear were to break while the vehicle was traveling at a high road speed, it would cause extremely rapid downshifts. After the truck was stopped you would find you had only first range forward or reverse range; you could never upshift to second range.

The governor valve receives its oil from the main oil pressure passage. Before this oil can reach the governor valve, it must flow through a governor filter. If the filter were plugged, then the governor could not get the correct oil pressure. This filter is often neglected during service work. There are two different types: a conical type or a barrel type (Figures 8-3 and 8-4).

Figure 8-3 An MT 643 four-speed transmission with a conical type governor filter.

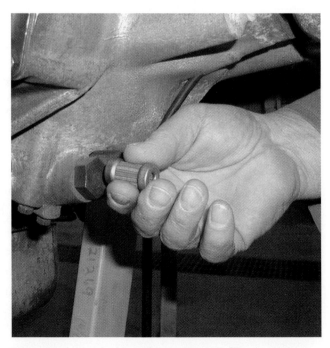

Figure 8-4 A typical barrel type governor filter located behind the large nut on an MT 653 five-speed transmission.

Torque Converter Torque Ratio

The torque converter changes the input speed and torque that goes into the transmission from the engine. It is located between the engine and the transmission (Figure 8-5).

Figure 8-5 Torque converter.

The proper torque converter torque ratio is selected after the information on the engine horsepower and torque, the rear axle ratio, the size of the tires, the weight of the vehicle; the terrain the truck will work in; and the number of transmission ranges (four- or five-speed) has been assembled. Stall torque ratio numbers can range from as high as 3.04:1 to as low as 1.82:1. What do these numbers mean?

Like any transmission ratio, the number is given as *n*: one (e.g., 3.04:1). This means that the engine torque is multiplied by 3.04 at stall (output stopped). In this case, if the engine could develop 400 foot pounds of torque (542 N·M), the maximum force the torque converter could output to the transmission would be 400 foot pounds times 3.04, or 1648 N·M foot pounds (5408 N·M). This would happen only if the engine were at full throttle and the truck were just starting to move, such as during a hard acceleration out of the fire hall. This is called a stall condition, and it happens only when the torque converter is at maximum vortex flow. This high vortex flow comes at a price, and that price is the generation of very high heat in the oil flowing inside the torque converter. This heat is very destructive to the oil and therefore also to the transmission if allowed to continue too long. The truck should have a transmission oil converter out temperature gauge. If the assembly plant did not include a transmission oil converter out temperature gauge, have one professionally installed.

The torque converter torque ratio is further multiplied by the transmission gear ratio, thus increasing torque on the U-joints and the fire pump transfer case. Ensure U-joint serviceability before conducting a stall test, or catastrophic drive-shaft failure may result.

Example: In first range an MT 643 has a forward ratio of 3.58:1, and when multiplied by the torque converter torque ratio of 3.04:1 it has a "stump-pulling" ratio of 10.88:1. An MT 653 deep reduction model using the same torque converter could produce a very low ratio of 24.42:1 in first range.

Each torque converter is matched to the truck. Any changes to the tire size, rear axle ratio, or engine may require a change in torque converter ratio and valve body adjustment. See your authorized Allison dealer for support.

CAUTION:
When you rebuild or replace a transmission, make sure to use a torque converter of the same ratio. The outside of a torque converter has few identifying marks, and it is very easy to mistake the ratios.

Engine RPM The engine rpm is very important for several reasons. When accelerating rapidly out of the fire hall, the transmission is designed to upshift to the next highest range at just below fully governed engine rpm (somewhat less than no load governed or high idle speed value given by some manufacturers).

Example: If the engine is intended to run at a maximum of 2000 rpm, then a full-throttle upshift on this model should occur at about 1800 rpm. Information on throttle upshift points is in the service manual: get a copy. If the engine is unable to reach 1800 rpm, then the transmission will never shift to a higher range when the throttle pedal is fully depressed.

I have actually encountered a well-meaning soul who reduced the truck's rpm because he honestly thought that he was going to make the engine last longer if it did not go as fast. The engine's maximum rated rpm must match the transmission's maximum rated rpm. This information is available on the engine and on the pump panel. Not only must these numbers be the same, but the engine must actually be able to attain its rated rpm. (See the stall test section later in this chapter for further information on the relationships among the transmission, torque converter, and the engine.)

Modulated Lockup When a four-speed transmission is in either second or third range (depending on the model; again this information is in the service manual), the torque converter will lock up and in effect go into a one-to-one condition as you travel down the road. This means that the torque converter will not change the engine torque and speed; this is done to improve engine fuel economy, but it also has an effect of engine braking as you come to a stop. With your foot off the throttle, the truck's engine will act as a brake to help stop the truck. A good driver can use this method of braking to increase the life of the service brakes. The modulated lockup's job is to hold the torque converter in this condition as the truck slows. In some models this engine braking can be held almost until the truck comes to a complete stop; in others it may be held until about 10 miles per hour. Not all models have a modulated lockup. The assembly number will indicate if the transmission has a modulated lockup. Refer to these numbers in the parts manual before you disassemble a transmission to repair a defective modulated lockup valve on a transmission that does not have one!

It is impossible to stress too much the importance of recording the SAM numbers of all vehicles in your fleet before the plates fall off. Any maintenance, repair, or diagnostic work will be almost impossible if you do not have this information. Store these numbers in a safe place for later use.

As described above, the transmission oil can get very hot, and this high heat is destructive to the oil. The normal operating temperature for transmission oil is between 160° F and 200° F (71°–93° C), and the

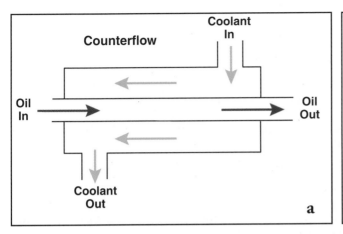 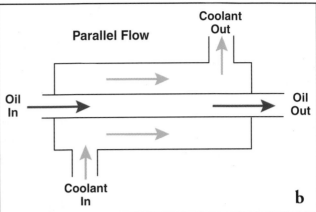

Figure 8-6 Two designs of transmission oil flow: counterflow (left), and parallel flow (right).

temperature should never exceed 300° F (149° C). The transmission oil cooler uses coolant from the radiator to remove this destructive heat from the oil. Normally, oil-flow routing follows a counterflow design as shown in Figure 8-6a. With this design of oil cooler, the oil and coolant must flow in opposite directions for the most efficient heat transfer. Figure 8-6b shows the less efficient parallel design that is sometimes used.

If the vehicle uses a transmission retarder, then an even larger capacity cooler must be specified when the truck is assembled. When active, the retarder uses the transmission oil to convert the truck's forward speed and energy to heat, just as the wheel brakes convert the forward motion of the truck into heat when they stop the truck. If this heat is not removed from the oil, the oil will very quickly deteriorate, becoming black or very dark instead of the red of normal Dexron oil. The oil may also have a burned smell. It will deposit a sticky varnish coating on the valves in the transmission, which will in turn cause the transmission to have erratic and unpredictable shifts both up and down. If this condition is allowed to continue, then the clutch packs may not get the correct pressure to engage fully; this will cause the clutch packs to burn when they slip under engine power. The burned clutch packs will release more contamination into the oil, deteriorating it further. Keeping the oil from overheating and changing the oil and filters as required can ensure long transmission life.

Example: fire etc. (formerly the Alberta Fire Training School) in Vermilion, Alberta, has 15 trucks with various models of Allison transmissions. Students drive these trucks very hard all spring, summer, and fall, and I suspect this use almost equals that of the busiest city department. These transmissions' oil and filters are changed once a year, and the school has never experienced any transmission problems that were not caused by the operators themselves.

CAUTION:

After a transmission rebuild, technicians often confuse the oil cooler lines and connect them to the cooler incorrectly. This will not cause an immediate overheating problem, but during the first trip with full throttle accelerations on a hot day, the transmission oil will begin to overheat.

Stall Test

A stall test should be conducted any time there is a concern about the performance of either the engine or the transmission. This test does not damage either of these components any more than a regular medical check-up damages a human being.

1. Before conducting a stall test, ensure that the transmission oil level is correct. An oil level that is either too high or too low will cause the oil to foam. The foamed oil contains air bubbles, which can cause incorrect test results.

2. Ensure that the engine and transmission are at operating temperature (between 160° F and 200° F [71–93° C]).

3. Place the truck in an open area, warn all people in the vicinity of the test, block the wheels, and apply *both* the parking brakes (spring) and the service brakes (air). Do not apply just the parking brakes because they engage only the rear axle.

4. Inspect the driveshaft and U-joints to confirm they can withstand the load that will be placed on them.

5. Make certain that the engine tachometer is accurate.

6. Place the transmission in the highest range possible; this means selecting D for drive. With models that use a DR, never select low range (D1), and with models that have a CR, never select reverse.

7. With the brakes firmly applied, increase the engine speed with the foot throttle to wide-open throttle (WOT). The engine should speed up until it reaches its correct stall speed; this rpm should be safely recorded.

8. The stall test should last no longer than 30 seconds (the rpm normally will stabilize in less than thirty seconds).

9. During the test, observe the transmission oil temperature on the dash. Never allow the temperature to rise above 300° F (149° C) during the test. Normally the oil's temperature rises very quickly.

Stall Speed RPM When your truck was manufactured, the builder took into account many factors: engine power, transmission ratio, rear axle gear ratio, tire size, top speed, gross vehicle weight (GVW), and the terrain over which the vehicle would have to operate. The manufacturer had to select the correct engine, torque converter, and transmission to allow the engine to produce near its peak torque at a specific rpm. This rpm is lower than the engine's maximum governed speed. For example, the stall speed of a common truck engine may be 1700 rpm with the maximum governed speed at 2300 rpm. The correct combination results in a truck that can pull away from a standing stop and accelerate to its top speed as fast as possible without damaging driveline components.

The job of multiplying engine torque is done inside the torque converter. The engine turns the torque converter impeller at whatever the engine speed is. The torque converter's action creates a flow of transmission oil that is pumped from the impeller to the torque converter turbine. The oil must now be redirected back to the torque converter impeller wheel; this is the stator's job. The stator is located between the impeller and the turbine wheel at the inner circle of the torque converter. During full throttle acceleration, the stator will lock (not be allowed to rotate). A one-way over-running clutch (also called a roller clutch) accomplishes the locking action, which is necessary for multiplying engine torque because of the redirection of the oil. If this clutch were to fail, engine torque would not multiply. For example, a normal stall speed of 1700 rpm would fall by approximately one-third, to 1200 rpm, if the overrunning clutch failed. The complaint would be that the engine had no power when accelerating. Some shops have taken large diesel engines apart and rebuilt them because of this no-power complaint. When reassembled, these engines still had no power because the problem actually was a defective over-running clutch in the stator.

Cool-Down Test At the end of the stall test, the transmission's oil will be very hot and must be cooled.

1. Slow the engine to idle speed and place the transmission in neutral. Accelerate the engine to between 1000 and 1400 rpm. You should now see the transmission oil temperature rapidly decrease. This part of the test is called the cool-down.

2. If the oil does not cool, then the oil cooler may be plugged or the one-way clutch in the stator could be stuck and may not be free-wheeling.

3. The cool-down test and its results are as important a diagnostic tool as the rpm stall test.

Many transmissions are ruined because of overheated oil, so the condition of the oil cooler is very important. As most transmission oil coolers use the engine coolant system to remove heat from the transmission's oil, any problem with the engine's coolant system will affect the transmission.

Interpreting Stall Test Results Depending upon whether they are above rated speed (also called a high stall speed) or lower than normal speed (also called a low stall speed), stall test results generally can help the vehicle technician isolate and identify specific transmission and engine problems.

Above Rated Speed Let's say that a normal stall speed for the truck being tested is 1700 rpm. If your results are more than 150 rpm above the normal rpm and the oil level was correct (remember to check the easy stuff first), then you most likely have a defective clutch pack. This could be because of low clutch-pack oil pressure (possibly a broken clutch-pack seal). Check the transmission operating main pressure and see if it is lower than normal.

This problem will not get better; transmissions rarely heal themselves. Avoid the use of transmission oil additives that are advertised to swell the clutch pack seals. These may be a quick fix, but they are not what you want for your fire truck. When in drive and at the curb conducting a stall test, remember that you have charged only the first and forward clutch packs, and therefore, a stall test tells you the condition of only those clutch packs. The clutch packs involved in the other forward ranges can be checked only when driving the truck. If you have a defective clutch pack in, say, third range, then you must look at the clutch packs involved in that range. In third range the clutch packs are forward and third clutch. If you know the forward clutch pack is good from the stall test, then the defective pack must be the third clutch pack.

How will you know you have a problem? When the transmission upshifts from a lower range to a higher range, you should see the engine rpm drop and then slowly climb again during a normal acceleration. If the rpm instead increases very quickly and the truck does not move forward at the same rate, then you have a flare. This flare is caused by a slipping clutch pack. For the transmission to be in any forward or reverse range, two clutch packs must be engaged.

Lower than Normal Speed If the stall speed is 150 rpm or more below the normal speed, the problem is most likely in the engine. If the engine is producing a lot of black exhaust smoke, suspect a plugged air filter or a defective fuel injector. If smoke is not excessive, then look at a plugged fuel filter. The cause could be a defective turbocharger or a crack in the air-to-air cooler in front of the radiator. It also could be that the engine just needs to have its exhaust valves, intake valves (if it has them), and injectors adjusted.

If the stall speed is substantially lower (30–50 percent below normal stall speed), a defective freewheeling stator will almost always be the problem. The complaint will be that the engine has no power, when in fact the torque converter is not multiplying the engine torque. Talk to the operators; they will most likely tell you that sometimes when they start off slowly (not too much throttle) the truck works well and at high speed in third or fourth range it works fine. Correct this right away; at this point it can be a fairly inexpensive repair. You may need only to install a new over-running clutch if no other damage has been done. If you allow this condition to continue, however, the one-way clutch may pile up debris inside the torque converter. This will cause the stator to stick. The torque converter will now multiply torque, but the transmission oil will overheat very quickly, and the transmission will be ruined. You will hear the operators say the transmission cured itself (be very worried if you ever hear that anything cured itself), but the next complaint will be that the transmission oil is very hot and smells burned.

The stall test should not be done as part of a daily or weekly check; you need to perform it only if you receive a complaint. You may, however, want to make stall testing part of an annual check. Obtaining the stall speed specification many years after the truck has been made can sometimes be difficult, so make sure you demand this information as part of the bid process and store it with the transmission's SAM number.

Fire Truck Transmission Applications

Mounting the Allison automatic transmission at midship necessitates modifying the transmission. Normally when it is placed in drive (D) at a curb, the transmission starts in first range and then upshifts to the next higher range as forward speed increases. When the forward speed is sufficient, the torque converter locks up (one-to-one ratio) and the transmission shifts into the higher ranges (third and fourth).

When the operator places the transmission into pump mode, two things must happen. The transmission oil has to obtain high range instantly. (This will be fourth range in the four-speed transmission and fifth range in the five-speed transmission. In both cases it will place the transmission in a direct range [one-to-one ratio].) At the same time the torque converter must also lock up. During pumping operations it is very important that the transmission does not try to shift out of this direct range or into any other ranges. The transmission must stay in this direct range at all times so that any increase in engine throttle will produce a direct and corresponding change in pump output. If you suspect a change in pump speed during pumping, then you will have to correlate the engine's speed (input) to the pump's speed (output). They must be the same; if not, then you have a problem. The transmission is slipping, the torque converter lock-up clutch is slipping, or the valves used to modify the transmission are defective.

Figure 8-7 shows the external hydraulic plumbing needed to accomplish this high-range lockup in a typical modification to an Allison transmission. The valves are electrical over hydraulic. That means a small amount of electrical power (12 volts and only a few amps) controls a flow of transmission oil.

Figure 8-7 External hydraulic circuit. (Courtesy of Allison Transmissions)

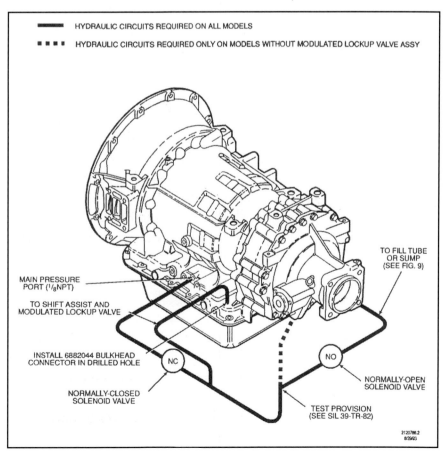

HYDRAULIC CIRCUITS REQUIRED ON ALL MODELS

HYDRAULIC CIRCUITS REQUIRED ONLY ON MODELS WITHOUT MODULATED LOCKUP VALVE ASSY

MAIN PRESSURE PORT (1/8 NPT)

TO SHIFT ASSIST AND MODULATED LOCKUP VALVE

INSTALL 6882044 BULKHEAD CONNECTOR IN DRILLED HOLE

NC

NORMALLY-CLOSED SOLENOID VALVE

TO FILL TUBE OR SUMP (SEE FIG. 9)

NO

NORMALLY-OPEN SOLENOID VALVE

TEST PROVISION (SEE SIL 39-TR-82)

There are two types of these valves, those that are normally closed (NC) and those that are normally open (NO). When electrical power is applied to an NC valve, the valve opens and allows oil to flow through it. Conversely, when electrical power is applied to an NO valve, the valve closes.

Any corrosion of the wires or their connectors can cause a voltage drop to these valves, and then they will malfunction. If they stick or become plugged with contaminated hydraulic oil, then the transmission either sticks in high range or fails to reach the high range necessary for pumping operations. Figure 8-8 illustrates plumbing for a typical Allison transmission that uses a modulated lockup valve; Figure 8-9 illustrates plumbing for a typical Allison that does not use a modulated lockup valve.

All other modifications needed to make these transmissions more useful to the fire service will be made to the valve body inside the transmission. These modifications are necessary to achieve both the lockup condition and the fourth range. The particular modifications required depend on the model and assembly numbers; consult the parts and repair manuals and your local Allison dealer for more information.

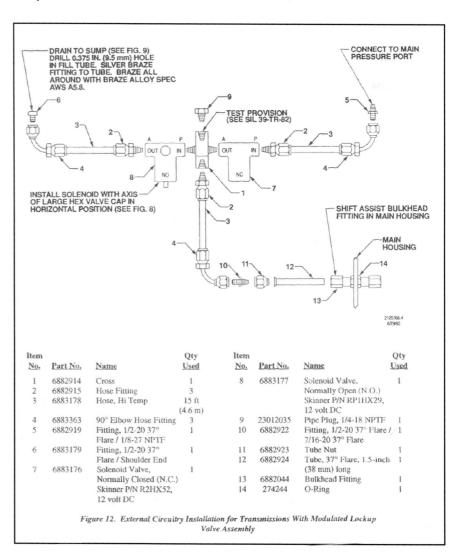

Figure 8-8 External circuitry installation for Allison transmission with modulated lockup valve assembly. (Courtesy of Allison Transmissions)

Item No.	Part No.	Name	Qty Used		Item No.	Part No.	Name	Qty Used
1	6882914	Cross	1		8	6883177	Solenoid Valve, Normally Open (N.O.) Skinner P/N RP1HX29, 12 volt DC	1
2	6882915	Hose Fitting	3					
3	6883178	Hose, Hi Temp	15 ft (4.6 m)					
4	6883363	90° Elbow Hose Fitting	3		9	23012035	Pipe Plug, 1/4-18 NPTF	1
5	6882919	Fitting, 1/2-20 37° Flare / 1/8-27 NPTF	1		10	6882922	Fitting, 1/2-20 37° Flare / 7/16-20 37° Flare	1
6	6883179	Fitting, 1/2-20 37° Flare / Shoulder End	1		11	6882923	Tube Nut	1
7	6883176	Solenoid Valve, Normally Closed (N.C.) Skinner P/N R2HX52, 12 volt DC	1		12	6882924	Tube, 37° Flare, 1.5-inch (38 mm) long	1
					13	6882044	Bulkhead Fitting	1
					14	274244	O-Ring	1

Figure 12. External Circuitry Installation for Transmissions With Modulated Lockup Valve Assembly.

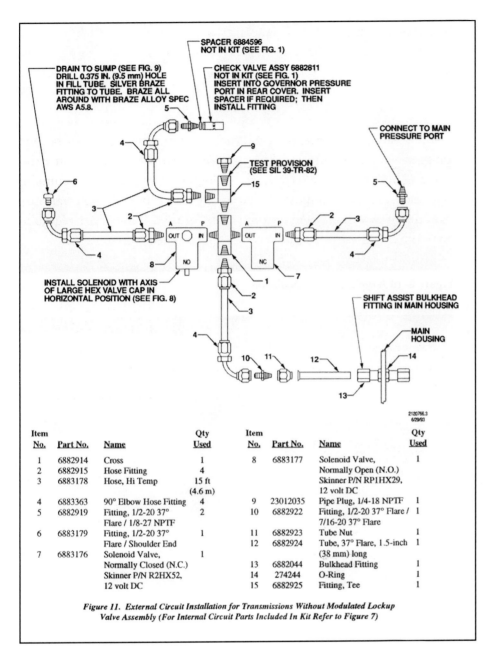

Figure 8-9 External circuit installation for Allison transmissions without modulated lockup valve assembly. (Courtesy of Allison Transmissions)

Item No.	Part No.	Name	Qty Used	Item No.	Part No.	Name	Qty Used
1	6882914	Cross	1	8	6883177	Solenoid Valve, Normally Open (N.O.) Skinner P/N RP1HX29, 12 volt DC	1
2	6882915	Hose Fitting	4				
3	6883178	Hose, Hi Temp	15 ft (4.6 m)				
4	6883363	90° Elbow Hose Fitting	4	9	23012035	Pipe Plug, 1/4-18 NPTF	1
5	6882919	Fitting, 1/2-20 37° Flare / 1/8-27 NPTF	2	10	6882922	Fitting, 1/2-20 37° Flare / 7/16-20 37° Flare	1
6	6883179	Fitting, 1/2-20 37° Flare / Shoulder End	1	11	6882923	Tube Nut	1
				12	6882924	Tube, 37° Flare, 1.5-inch (38 mm) long	1
7	6883176	Solenoid Valve, Normally Closed (N.C.) Skinner P/N R2HX52, 12 volt DC	1	13	6882044	Bulkhead Fitting	1
				14	274244	O-Ring	1
				15	6882925	Fitting, Tee	1

Figure 11. External Circuit Installation for Transmissions Without Modulated Lockup Valve Assembly (For Internal Circuit Parts Included In Kit Refer to Figure 7)

Figure 8-10 depicts a typical valve installation. The electrical connection is poorly made; never use this type of butt connector unless you have the time to come back later and replace it properly. The correct way to make electrical connections is to strip the insulation first with the correctly sized wire stripper, not your jack knife. Then crimp on a connector without the plastic covering. Next, solder the connection, and finally waterproof the connection to keep out any corrosion. Remember that the crimp provides only the mechanical connection for the wire; the solder flows between the wires to provide the electrical connection. Many different methods are available to waterproof an electrical connection. Shrink tubing that is heated and then allowed to shrink around the connection is very popular. Self-vulcanizing electrical tape is also very good.

Figure 8-10 A typical valve installation.

Figure 8-11 Breather on an automatic transmission.

Breather

The transmission will have a breather on the top of the transmission case (Figure 8-11). If this breather becomes plugged, then pressure can build in the transmission case and push oil out around the input seals, output shaft seals, or fill tube. As part of routine maintenance, spin this breather with your hand to ensure it is not plugged.

Electronically Controlled Transmissions

Electronically controlled transmissions offer unique opportunities but also present unique problems. The Allison transmissions you are most likely to encounter, the MD and HD models, belong to Allison's family of WT transmissions (WT stands for World Transmissions). Although this family of transmissions has basically the same three planetary gear sets as the four-speed MT and HT models, they can obtain a maximum of six forward speeds. Fourth range in the new WT family is the direct one-to-one ratio that will be used in pump mode. The torque converter also will be in lockup during pump mode (this is similar to the older automatic transmissions of the MT and HT families).

While the WT can have six forward speeds in normal configuration (with fifth and sixth speeds being overdrives), not all of these ranges may be available. When they specify the truck's transmission, the designers take into consideration the weight (GVW), the engine horsepower, axle ratio, top speed, and other concerns to determine whether it is safe to allow the transmission to be able to obtain these high-speed ranges. If they deem these ranges dangerous, they instruct (burn) the transmission's electronic control unit (ECU) chip not to allow them. It is not possible to change this, and you should not even try. If the ECU fails and a new ECU is needed, source only from your Allison dealer to ensure an exact match. Many fleets have two WT transmissions that are identical except that one has five forward ranges and the other has six. In any case, they will both pump water in fourth range.

The SAM number plate is not as critical on an electronic transmission as on the MT and HT units because the information it contains is also available from the ECU by way of an electronic reader. One very popular and easy-to-use reader is the Pro-Link. The Pro-Link 9000 is a microprocessor-based handheld tester capable of protocols J1708, J1939, and ISO-9141. By inserting the multiprotocol cartridge into the back of the Pro-Link and one of many available application cards, you will have access to many different systems (Figure 8-12).

Figure 8-12 Pro-Link 9000. From left to right are the Allison transmission card, the Detroit diesel engine card, the Navistar engine card, and the Meritor Wabco antilock brake card. Many other cards are available. The three most popular connectors are shown.

These units can retrieve trouble codes and transmission operating information. Trouble codes (and transmission-oil level information on some models) are also available from the shift selector display. In either case a maximum of five trouble codes can be stored. By retrieving these codes, technicians will have an excellent indication of the area of trouble. Some codes will trigger a do-not-shift (DNS) light on the dash. When this occurs the transmission will ignore the Any Range selection from the shift selector and will not upshift or downshift, depending on what type of failure has caused the code. Not all codes will cause a DNS condition; in many cases the operator will not know there is a problem. If the condition that set the code does not reoccur and therefore becomes inactive, the code will disappear after twenty-five key starts. Your maintenance program's regular inspection procedure should include retrieving these codes.

Inputs and Outputs

Like any computer controlled system, the WT transmission needs both inputs and outputs (Figure 8-13). The inputs, which may be sensors or on/off switches, provide the computer with information about conditions onboard the truck. Sensors may include speed sensors (giving an AC frequency voltage), oil temperature sensors, and throttle position sensors. On/off switches include shift selectors (what range you want), C-3 clutch pressure, PTO enable, engine brake or transmission retarder, antilock brake response, and of course, the most important for us, fire truck mode. The outputs are what the computer tells the transmission to do. Such outputs could be clutch engagement, torque converter lockup, and indicators and lights such as the DNS light and the range, retarder, and PTO indicators.

Electronically controlled transmissions can adapt to (learn) different operating conditions. The transmission will always try to give the smoothest shift possible, and to do so it will fill and dump clutch packs at a suitable rate. If the engine is being driven for a period of time in wide-open throttle mode (WOT), then the transmission will charge the clutch

Figure 8-13 Inputs and outputs for an electronic transmission. (Courtesy of Allison Transmissions)

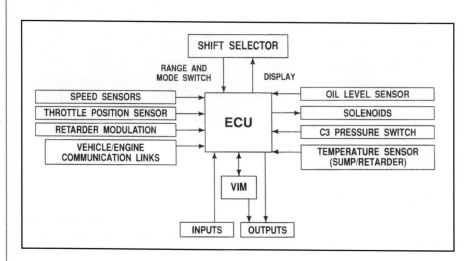

packs very aggressively to ensure that they do not slip and burn. This will give that operator a very solid and firm shift. If a different driver with a less aggressive driving style then used this same truck, that driver might complain of harsh shifts. The transmission would have adapted to the first driver's driving style and will take some time to learn the second driver's less aggressive style.

An electronically controlled transmission's ECU must obtain full system voltage; low voltage to the ECU caused by corroded or faulty connections will stop the transmission from shifting. Most often the problem causing excessive voltage drop is in the ground circuits. A computer main code of 13 with subcodes of 12, 13, or 23 indicates an electrical problem. If the transmission is fitted to a nonelectronic engine, you will need to use a throttle position sensor (TPS). This sensor tells the transmission's ECU where the driver has the foot throttle (pedal). If the throttle is at or near full throttle, the transmission will wait longer to upshift, and when it does upshift, it will do so aggressively. If the throttle is very light, the transmission will upshift sooner and more smoothly. The TPS on the WT fulfills the duty of the modulator on the AT/MT/HT mechanical transmissions.

Failure of the TPS will result in a computer main code of 21 with a subcode of 12 or 23. If an electronic engine powers your transmission, the transmission does not need a separate TPS. The TPS that the electronic engine uses will be able to communicate with the transmission through the engine's data link. This transmission uses three speed sensors: one that measures the engine speed from the torque converter, a second that measures the speed of the turbine shaft, and a third that measures the speed of the output shaft to the rear axle. A failure in these sensors will result in a computer main code of 22 with a subcode of 14 for a failed engine speed sensor, a subcode of 15 for a failed turbine speed sensor, and a subcode of 16 for a failed output speed sensor. When in pump mode for a fire truck, all three sensors should read the same speed indicating both torque converter lockup and fourth range (direct).

Transmission sump temperature will appear as a main code of 24 with a subcode of 12 if the temperature of the oil is too low or with a subcode of 23 if the temperature is too high. At a temperature of –25° F (–32° C) the transmission is too cold to operate, and you will need to warm the transmission before it will move the truck. A Pro-Link temperature reading of –49° F (–45° C) indicates a temperature sensor failure. A temperature sensor failure will give a main code of 33 with subcode of 23 for a failed high sensor or a subcode of 12 for a failed low sensor.

Lubrication and Filter Requirements

As with the older automatic transmissions, routine oil and filter changes can greatly extend the service life of electronically controlled transmissions. Both hydraulic and electronic transmissions use a Dexron 3 automatic transmission fluid (ATF). For off-highway applications, use a fluid that meets the requirements of C-4 oil. Off-highway is considered aggressive driving; hard, full-throttle acceleration; high-temperature operation; and the use of a transmission retarder. If this sounds like your department, then switch to C-4 oil.

> # CAUTION:
> **Many C-4 oils are also motor oils, but not all motor oils meet the requirements of C-4 transmission oil. Switching to C-4 oil might allow you to use the same type of oil in both your engine and your transmission.**

An optional oil-level sensor is available on the World Transmission. This sensor can indicate if the transmission is as much as four quarts high or four quarts low. While it does not replace the dipstick for the most accurate check of transmission-oil level, it is very handy if getting to the dipstick on the transmission is difficult. If your operator uses a dirty rag to wipe the dipstick or forgets to replace the dipstick properly, it may be best to use the oil level sensor. For a correct reading, the transmission must be warmed to between 140° F (60° C) and 220° F (104° C); the engine must be at an idle speed of less than 1000 rpm; and the vehicle must have been stationary for at least two minutes to allow the transmission oil to settle.

A computer main code of 14 with subcodes of 12 or 23 indicates an oil sensor problem.

NOTE: Dexron 3 replaced Dexron 2, and C-4 oil replaced C-3. Use of these new oils in an older transmission is recommended.

Transmission oil and filters should be changed after the first 5,000 miles (8,000 km) and then every 25,000 miles (40,000 km) or 18 months. These change intervals are based on highway applications and may not reflect your department's driving habits; therefore, oil analysis should be conducted to confirm this oil change interval. Change transmission oil filters at every oil change. Unlike engine oil filters, the Allison oil filter does not use an oil bypass valve. If the oil filters were to plug up with debris, the transmission components downstream from the filter would not receive enough oil flow. Thus, you should use the original Allison filters because of their ability to hold dirt. Use of a non-original-equipment-manufacturer (OEM) filter that may have a finer mesh may result in cleaner oil, but it may also result in a plugged filter and an oil-starved transmission.

Fire Truck Mode for the WT Transmission
The diagrams in Figures 8-14 and 8-15 show typical electrical connections. They are from the WTEC III Electronic Controls Troubleshooting Manual.

APPENDIX P — INPUT/OUTPUT FUNCTIONS

WARNING!	These schematics show the intended use of the specified controls features which have been validated in the configuration shown. Any miswiring or use of these features which differs from that shown could result in damage to equipment or property, personal injury, or loss of life. ALLISON TRANSMISSION IS NOT LIABLE FOR THE CONSEQUENCES ASSOCIATED WITH MISWIRING OR UNINTENDED USE OF THESE FEATURES.

J. FIRE TRUCK PUMP MODE

USES: Facilitates engagement of split shaft PTO and shifts transmission to fourth range lockup.

VARIABLES TO SPECIFY: None

VOCATIONS: Fire Truck Pumpers

WARNING!	If this function is enabled in the shift calibration, the function MUST be integrated into the vehicle wiring. If the function is available in the shift calibration but will not be used in the vehicle, it MUST be disabled in the calibration.

SYSTEM OPERATION

OPERATOR ACTION — System Response

TO ENGAGE:

1. *SELECT NEUTRAL* — Transmission shifts to Neutral.

2. *APPLY PARKING BRAKE* — None

3. *SELECT PUMP* — Turns on "Pump Mode Requested" light. Stops output shaft rotation. When split-shaft engages, PPE signal and "Pump Engaged" light are turned on. Transmission output unlocks.

4. *SELECT DRIVE* — Transmission shifts to fourth lockup. "OK To Pump" light is turned on.

TO DISENGAGE:

1. *SELECT NEUTRAL* — Transmission shifts to Neutral if output rpm < 1000.

2. *SELECT ROAD MODE* — Stops output shaft rotation. PTO disengages. Transmission shifts back to Neutral. If the output shaft rotation continues, press the momentary transmission brake switch before selecting road mode. This will cause the transmission output shaft to stop if transmission is in neutral and shaft rotation is less than 100 rpm.

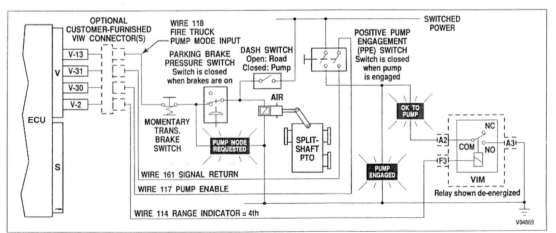

Figure P–16. Fire Truck Pump Mode

Copyright © 1999 General Motors Corp.

Figure 8-14 Electrical connections for fire truck pump mode. (Courtesy of Allison Transmissions)

APPENDIX P — INPUT/OUTPUT FUNCTIONS

> **WARNING!** These schematics show the intended use of the specified controls features which have been validated in the configuration shown. Any miswiring or use of these features which differs from that shown could result in damage to equipment or property, personal injury, or loss of life. ALLISON TRANSMISSION IS NOT LIABLE FOR THE CONSEQUENCES ASSOCIATED WITH MISWIRING OR UNINTENDED USE OF THESE FEATURES.

J. FIRE TRUCK PUMP MODE (OPTIONAL)

USES: Facilitates engagement of split shaft PTO and shifts transmission to fourth range lockup.

VARIABLES TO SPECIFY: None

VOCATIONS: Fire Truck Pumpers

> **WARNING!** If this function is enabled in the shift calibration, the function MUST be integrated into the vehicle wiring. If the function is available in the shift calibration but will not be used in the vehicle, it MUST be disabled in the calibration.

SYSTEM OPERATION

OPERATOR ACTION — System Response

TO ENGAGE:

1. SELECT NEUTRAL — Transmission shifts to Neutral.

2. APPLY PARKING BRAKE — None

3. SELECT PUMP — Turns on "Pump Mode Requested" light. Turns on both input signals to ECU (wires 117 and 118) which activates "fire truck" mode. When split-shaft shifts, "Pump Engaged" light is turned on.

4. SELECT DRIVE — Transmission shifts to fourth lockup. "OK To Pump" light is turned on.

TO DISENGAGE:

1. SELECT NEUTRAL — Transmission shifts to Neutral if output rpm < 1000.

2. SELECT ROAD MODE — PTO disengages. If the output shaft rotation continues, press the momentary transmission brake switch before selecting road mode. This will cause the transmission output shaft to stop if transmission is in neutral and shaft rotation is less than 100 rpm.

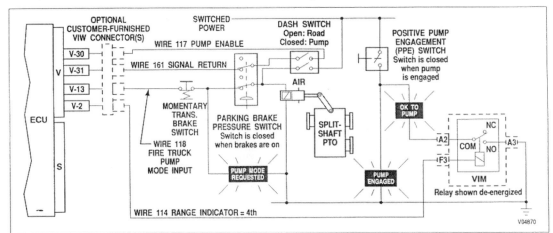

Figure P–17. Fire Truck Pump Mode (Optional)

Copyright © 1999 General Motors Corp.

Figure 8-15 Optional electrical connections for fire truck pump mode. (Courtesy of Allison Transmissions)

Troubleshooting the WT Transmission

When troubleshooting the rare electrical/electronic problems that can occur with electronically controlled transmissions, do not jump to conclusions. Apply solid electrical theory (Ohm's Law), use high-quality accurate meters, get the troubleshooting service and parts manuals for the transmission you have, and if at all possible, attend a factory-authorized school at your local Allison dealer. Bad electrical connections or incorrectly installed after-market accessories cause most problems.

Summary

The Allison Transmission is very reliable and very adaptable to the conditions experienced during fire-fighting operations. Its ease of operation allows firefighters to concentrate on more critical emergency functions than changing ranges. To achieve its full potential serviceable life, follow a regular and thorough maintenance schedule that addresses its unique use within the fire service.

Chapter 8 Review Questions

Short Answer

Write your answers to the following questions on the blank lines provided.

1. List three inputs of the mechanical automatic hydraulic transmission.

   _____

   _____

   _____

2. Is a PTO option available on the MT transmission?

   _____

3. Explain the difference between a deep ratio reduction and a close ratio reduction transmission.

   _____

   _____

   _____

4. Name the two devices used to clean the oil in an HT 700 model transmission and state their locations.

   _____

   _____

5. Are cork gaskets recommended for automatic hydraulic transmissions?

   _____

6. Explain the purpose of the SAM plate.

   _____

   _____

7. Where is the SAM plate located?

   _____

8. What is the job of the transmission governor?

   _____

   _____

9. How does the transmission governor accomplish its job?

   _____

   _____

   _____

   _____

   _____

10. What method is used to identify governor valves?

_____

11. The operator's foot varies the modulator valve. True or False?

_____

12. When does high vortex flow occur in a torque converter?

_____

_____

_____

13. Does high vortex flow create high heat in the transmission oil?

_____

14. Do torque converters multiply torque, rpm, or both?

_____

15. Will changing tire size or rear axle ratio affect transmission performance?

_____

_____

16. Explain how to determine whether a mechanical automatic hydraulic transmission has a modulated lock-up valve.

_____

17. Hydraulic transmission retarders can be used to slow the vehicle by converting mechanical energy to heat. How is this destructive heat removed?

_____

_____

18. Describe the effects of overheating the transmission oil.

_____

_____

_____

_____

19. Why is the cool-down test such an important component of a stall test?

_____

_____

20. If engine stall speed is significantly lower (33 percent) than specified, where is the problem most likely located?

_____

_____

21. Is the stall test destructive to the transmission?

_____

_____

22. In a fire truck transmission application, what conditions must be met for the fire pump to operate?

_____

_____

23. How are the conditions necessary for fire-pump operation accomplished on mechanical transmissions?

_____

_____

24. How are the conditions necessary for fire-pump operation accomplished on electronic transmissions?

_____

25. Electronically controlled transmissions use high voltages and high amperages. True or false?

_____

26. In what range is direct drive normally found in a six-speed electronically controlled transmission?

_____

27. Why is the SAM information plate not as critical with electronically controlled transmissions?

_____

_____

28. When a problem occurs in the electronically controlled transmission, will the DNS light always illuminate?

_____

29. At present how many trouble codes can the electronic transmission store?

_____

30. What mechanism does the electronic transmission use to ensure long operational life?

_____

31. The electronic transmission uses inputs and outputs. Are speed sensors inputs or outputs?

_____

Review Question Answers

Review Question Answers

Chapter 1

Short Answer

1. NFPA standards are minimums.

2. Are we satisfied with the present level of maintenance? Are we satisfied with the costs? Do we predict that the average age of our front line fleet will decrease? Do we have a program, plan, and money to train new technicians?

3. False. It was designed for big city departments *and small rural departments.*

4. The tasks of a Level I EVT are to inspect, conduct maintenance, and perform operation checks. A Level II EVT must be able to do everything that a Level I can do plus conduct service tests and make repairs.

5. Yes. EVTs have specialized knowledge that can be vital to selecting the most appropriate apparatus for any particular department.

6. Because of the unique use of the fire apparatus, the manuals written for over-the-road trucks do not apply. Cold-engine startups, wide-open throttle, and extreme heat are just a few of the differences between the use of fire apparatus and that of normal over-the-road trucks.

7. Because each department has different people, equipment, terrain, and weather conditions, it would be impossible to have a one-size-fits-all program. This does not mean you should give up, just make your own schedule.

8. Crisis, preventive, and predictive maintenance.

9. Crisis.

10. Preventive maintenance is time based, and it may not work for your fire department because it may be based on over-the-road truck uses.

11. Proactive.

12. Predictive.

13. By recording these amounts and breaking them down, you can begin to find problem areas and make improvements.

14. Same as 13.

15. Detailed record keeping is important for two reasons. Firstly, if you cannot measure it you cannot improve it; secondly, your records will be used in a court of law in the event of an accident.

16. NFPA 1901, NFPA 1911, and NFPA 1500, plus the FMVSS or the CVMSS.

17. Commercial driver's license.

18. So the piece being ground will not get caught between the wheel and the rest, which would cause the piece to fly off dangerously.

19. If the piece of machinery turns at the same speed or a multiple of the same speed at which the light is flickering, then the rotating machine will appear to be stopped.

20. Grade 5, grade 8, grade 12.

21. If air were allowed to get to the rags, spontaneous combustion could occur.

22. Workplace Hazardous Materials Information System.

23. Chemical composition, protective clothing requirements, cleanup procedures, and any other precautionary information on the substance.

24. No.

25. The worker is responsible for knowing about WHMIS and material data sheets.

Multiple Choice

1. a

2. b

3. a

4. d

5. c

6. d

7. d

8. d

9. a

10. d

Chapter 2

True or False

1. False. When a pumping apparatus conforms to NFPA 1901, it is said to be *standard* apparatus.

2. False. The applicable standard is *NFPA 1901*.

3. True. Aerial ladders are under stress both when the tips are supported and when they are not.

4. True.

5. False. All *elevating platform* apparatus require two control stations. The aerial ladder apparatus *does not*.

6. True.

Short Answer

1. The minimum rated discharge capacity for a standard pumper, per NFPA 1901, *Standard on Automotive Fire Apparatus*, is 500 gpm (2,000 L/min).

2. A triple-combination pumper must have a water tank, hose bed, and fire pump.

3. The main difference is their size and, therefore, the capacity of their pumps and the amount of equipment they carry.

4. A truck's maximum weight is called the gross vehicle weight (GVW).

5. This operation is called a shuttle.

6. Aerial ladder apparatus

   Elevating platform apparatus

7. Any three of the following:

   Auxiliary engine

   Cross-mounted engine

   Power take-off

   Midship transfer drive

8. Cross-mounted engine drive. It consists of an auxiliary-engine-driven pump that is mounted across the apparatus.

9. Brush trucks, minipumpers, and mobile water supply apparatus

Multiple Choice

1. a

2. c

Chapter 3

1. Flow pressure

2. Water is not easily broken down; water is an effective coolant; water has very little adhesive quality; water cannot be pulled or lifted; water can be pushed; water is nearly incompressible; water stores very little pressure.

3. Atmospheric pressure, static pressure, residual pressure, head pressure, normal operating pressure, flow pressure

4. Naturally, chemically, mechanically

5. The most common type of fire pump in use today is the centrifugal pump.

6. Impeller, case

7. *Positive displacement pump:* A positive action takes place, forcing all of the water and air out of the pump body each time one cycle of the operation is completed.

   Centrifugal pump: Centrifugal force creates the water velocity needed to attain the required pump discharge pressure.

8. Piston pump, gear pump

9. Positive displacement pumps are used in small capacity, high-pressure fire-fighting applications; as priming pumps during drafting operations; and as booster pumps.

10. It is not restricted to pumping a specific amount of water with each revolution of the pump.

11. No. The centrifugal pump is incapable of pumping air; it relies on the water's velocity to move it through the system.

12. Atmospheric pressure

13. Flow, pressure

14. High pressure

15. High flow

16. Pumping over very long distances using the high pressure to overcome friction in the hose length and pumping to very great heights

17. Transfer valve or changeover valve

18. Any two: electrical, vacuum, or air pressure

19. Intake valves must not open or close faster than _three_ seconds. They must not open or close slower than _ten_ seconds.

20. Water hammer

21. An intake valve exhausts water to the ground, and a pump relief valve exhausts water to the intake side of the fire pump.

22. To ensure the valve's position does not inadvertently open or close during use. This could create a dangerous situation for the firefighter.

23. Front mount, midship, and rear mount

24. Attempting to pump water faster than it can enter the eye of the pump

25. Open and flush pump drains, clean suction inlet strainers, check oil levels, reset governor or relief valves, close drain valves, flush systems that have pumped either salt or corrosive water, refill booster tank, clear air from the pump

26. The cooling and lubrication of the pump are very dependent upon the water. Running the centrifugal pump without water can overheat the pump and cause serious damage.

27. Power source, fuel levels, batteries, crankcase oil, air cleaner, oil filter, bearings, stuffing boxes, suction inlet, discharge outlet, gauges

28. Annually

29. Less water is needed to put out the fire, and clean-up after the fire is easier.

30. 0.1 to 1.0 percent

31. Flush the system after each use with plain water. Never mix foam agents (Class A with Class B). Do not allow air to contact the foam concentrate in the foam storage tanks.

32. The normal maximum outlet pressure of an eductor-type foam system generally can not exceed 130 psi (896 kPa). This could make delivery of the water and foam difficult if long distances or great heights have to be overcome.

33. True

34. False. The foam flows through the pump. Excess foam could cause pump cavitation.

35. True

36. Less foam concentrate is needed, and the fire hose is lighter and therefore easier to handle.

Chapter 4

True or False

1. False. Fire pumps are used more often on pumpers than on aerial apparatus.

2. False. Starting the engine unnecessarily is of no benefit to the engine and may, in fact, do harm.

3. True

4. True

Short Answer

1. Maintenance implies keeping something in a state of readiness and good repair. Repair means to fix something that has become damaged.

2. Any five of the following: check crankcase oil levels, check water level of radiator, check batteries, check visible and audible warning signals, check fuel levels, check water tank level, check tires for inflation as well as wear, operate foot pedal to check brakes, clean windows, wash and dry vehicle, operate changeover valves on two-stage pumps.

3. Any five of the following: check transmission oil level, check differential oil level, check power steering fluid level, check hydraulic brake system, check master cylinder brake fluid level, check for leaks in air system, bleed moisture from air brake tanks, check fan belt and generator or alternator belts, check battery terminals and cables, operate valves in cooling system, check drains and hose connections for security, check drive shaft and universal joints, start motor to observe oil pressure and idling speed, clean under chassis, clean the engine and electrical motors, check for loose nuts, studs, and pins.

4. a. Identify the positive and negative grounds of the battery.

   b. Attach the red cable to the positive terminal.

   c. Attach the black cable to the negative post.

   d. Connect the charger to a reliable power source.

   e. Set the desired battery charging voltage and charging rate, where equipped.

   f. Reverse the procedure to disconnect the charging cables.

5. Any three of the following: rotating lights, hose rewind reel, windshield wipers, apparatus controls, heater and defrost fan.

Chapter 5

1. Friction reduction, heat removal, contamination suspension, sealing

2. The compound that is an additive in motor oil that allows the oil to pour at low temperatures and protects engine parts at high temperatures is a polymer.

3. The four popular grades of multigrade oils are 0W-40, 10W-30, 0W-20, and 15W-40.

4. Yes; it is a common misconception that 0W-40 is a low-temperature oil only.

5. Oil

6. Grease

7. The grade 5 is thicker.

8. True. Too much grease will cause the bearing to overheat; the grease will melt; and the bearing will be ruined.

9. 21/17/13

10. The amount of contamination will double for every range number increase.

11. These particles have the same effect as sandblasting.

12. 18/16/13

13. Yes. The oil might come in a reusable 50-gallon (205 L) drum that has not been thoroughly cleaned; changing hand pumps from one barrel to another or using dirty buckets or pails will also contaminate the oil before it ever gets to the apparatus.

14. Nominal, absolute, and beta ratio. Beta ratio most closely indicates a filter's contamination trapping ability.

15. Beta ratio of 20; efficiency rating of 95 percent

16. When a surge of oil hits a filter, the filter's pleats can bunch up reducing its dirt-holding ability. The filter then flexes back to normal after the oil surge passes. This relaxing of the pleats releases dirt into the oil.

17. One manufacturer has avoided this bunching problem by placing fiberglass wraps around the pleats and a resin bead between the pleats.

18. To establish a proactive maintenance program

    To confirm a preventive/proactive maintenance program

    To meet warranty requirements

    To retain trade-in/resale value

19. Aluminum (Al): blower, camshaft bearings, turbocharger bearings, crankshaft bearings, pistons, oil pump bushing

    Copper (Cu): oil cooler, valve train bushings, camshaft bushings, thrust washers, wrist pin piston

    Chromium (Cr): piston rings, tapered or roller bearings, exhaust valves, liners

    Iron (Fe): cylinder liners, pistons, camshafts, crankshafts, rocker arms, valve bridges, and camshaft followers

    Silver (Ag): bearings, solder

    Tin (Sn): overlay on some pistons and bearings, bushings

    Lead (Pb): bearings, solder, octane improver (gasoline engines), and oil additive

20. Zinc, phosphorous, boron, calcium, lead, sodium, molybdenum, and barium

21. Spectrographic analysis can now measure particles only up to the 8-micron size.

22. As engine oil becomes polluted with soot, its viscosity increases.

23. Long periods of idling can cause the engine to run cold; this would lead to incomplete combustion and dilution of the oil by unburned diesel fuel.

24. Heat analysis could be used to show an engine cylinder that is running cold because of incomplete combustion.

25. As a bearing in the gearbox begins to fail it can spall off small pieces of the bearing material; this will make the bearing's surfaces uneven and will cause the bearing to vibrate and emit a different noise than normal.

26. Any three of the following:

    Pitted liners

Failed oil coolers. Aluminum corrosion products will stop the flow of coolant through the oil cooler causing severe ring and bearing wear due to improperly cooled engine oil.

Failed radiators

Extreme aluminum corrosion

Abnormal water pump and head gasket failure

Iron destruction caused by copper plating onto the iron components

Abnormal rusting of the cab and other sections of the equipment

27. The multimeter volt meter leads must be long enough to reach from the coolant to the ground side of the battery.

28. The proper meter lead is placed to the ground side of the battery, negative to negative or positive to positive, and the other lead is placed first in the coolant, then touching the outside of the engine block, and finally touching the top radiator tank. AC and DC readings are taken at each position.

29. To prevent a current generated by the rear-end differentials from traveling up the driveshaft and grounding through the engine coolant or the transmission coolant the rear ends and transmissions should be bonded and grounding straps installed to connect the rear ends to the ground (earth).

30. If the coolant shows an electrical problem with all the equipment turned on, turn off one system at a time until you finally turn off the system that stops the electrical current. Through this systematic process, you can isolate and detect the electrical system causing the problem.

31. The electrical current will destroy the iron protecting chemicals in a properly formulated coolant.

Chapter 6

1. Electron pump

2. Voltage

3. The R terminal

4. 14 volts

5. 10 percent

6. 5,000 rpm

7. Down

8. Synthetic

9. Clean

10. Heat

11. Brushless

12. 33 percent

13. 66 percent

14. Oscilloscope, DC ammeter, or an AC voltmeter

15. Piston

16. 17/15/13

17. Voltmeter, ammeter, frequency meter, and hour meter

18. An internal combustion engine produces a very deadly gas called carbon monoxide in its exhaust; the breathing air compressor must not take this gas in as it can kill anyone who breathes it.

19. A size 2 wire

20. 0.2 volts DC

21. En route

22. High-cycle

23. 25 amps at a minimum of 10.5 volts

24. In the event of an alternator failure, many electronic engines and transmissions will need at least 10.2 volts to remain operational.

Chapter 7

1. The engine's ability would be reduced because in warm air the oxygen molecules are farther apart and air that is moist contains water, which does not aid the combustion process.

2. The pressure of the atmosphere drops 0.5 psi (34.5 kPa) for every 1,000-foot (304.8 m) increase in altitude. Therefore, at 5,000 feet (1,524 m), the gauge should read 2.5 psi (17.2 kPa) less than at sea level, or 12.2 psi (84.1 kPa).

3. The decreased air pressure means less air pressure is available to push air into the engine's cylinders, which causes a drop in available horsepower.

4. Pumps make water flow.

5. Resistance to flow causes pressure.

6. When testing a fire pump that has a rated capacity of 1,250 gallons per minute, (4,732 L/min) the maximum suction hose size is 6 inches (150 mm), with one pump inlet allowed and a maximum lift of no more than 10 feet (3 m).

7. Residual pressure is the pressure that remains in the system when water is flowing. Static pressure is the water pressure at the hydrant when no water is flowing.

8. Insufficient water supply.

9. Water outlet pressure does not rise in correspondence to increased engine throttle. A noise that sounds like you are pumping gravel may come from the fire pump. (Note: you most likely would not hear this noise.)

10. Output pressure of 190 psi (1,310 kPa) minus hydrant residual pressure of 30 psi (207 kPa) would yield a net pump pressure of 160 psi (1,103 kPa).

11. The FEL is 8.4 feet (1.6 m); the static lift is 10 feet (3 m); and the dynamic lift is the sum of FEL and static lift, 18.4 feet (4.6 m).

12. A transducer is any device that changes one form of energy to another. An example is the paddle wheel flow sensor, which converts water pressure to voltage.

13. Electronic diesel engines produce less pollution than mechanical engines when operating at high altitudes.

14. False. Unless the engine has been specifically modified, its efficiency will decrease as elevation increases.

15. Input signals from the flow sensors or pressure transducers are sent to the central processing unit (CPU) in the electronic pump governor's computer, and then the CPU sends output signals to various devices according to a programmed set of parameters. Thus the CPU can automatically maintain the pressure set by the operator.

16. The voltages and currents used in both inputs and outputs are very small, in most cases in the millivolt and milliamp range. It is very important that electrical connections be clean and tight to ensure a good electrical connection that will have a low voltage drop. High voltage drops caused by corroded connections will cause false signals and readings.

17. This recalibration is necessary whenever an alternator pulley combination is changed on the engine's alternator as the input signal for engine rpm usually comes from the R terminal on the alternator.

18. The interlock prevents the governor from operating unless the pump is engaged. If the engine speed is increased from idle and the transmission is in a forward gear, the brakes most likely will be unable to hold the vehicle. This results in a very dangerous situation.

19. Manufacturer's test, certification test, hydrostatic test, and acceptance test.

20. A service test should be performed whenever a repair has been done to the truck that could affect the performance of the fire pump.

21. The minimum pressure at which a preservice hydrostatic test must be performed is 250 psi (1,725 kPa).

22. Three minutes

23. An independent certification organization should perform the certification tests because this protects both the manufacturer and the buyer.

24. All engine driven accessories must remain connected during the certification tests to simulate an actual firefight.

25. 2 hours

26. 150 psi (1,035 kPa) net pump pressure

27. Adding engine fuel is permissible during the 30-minute, 70 percent test and during the 30-minute, 50 percent test, but not during the 2-hour, 100 percent test.

28. 30 minutes for the 70 percent test and 30 minutes for the 50 percent test

29. *Pump capacity:* 750 gpm (2,850 L/min) or larger; *pressure:* 165 psi (1,318 kPa); *duration:* 10 minutes; *flow requirements:* the pump must produce the same water flow as for the 100 percent test.

30. The pressure-control test must be conducted from draft.

31. 30 psi (207 kPa)

32. For pumps of 1,250 gpm (4,732 L/min) or less, the vacuum primer should reach the required vacuum in no more than 30 seconds. For pumps of 1,500 gpm (5,678 L/min) or more, the test limit is 45 seconds.

33. From sea level to 2,000 feet (610 kPa) above sea level, the vacuum must be 22 inches of Hg (64.34 kPa).

34. 5 minutes

35. When with the fire pump turning, water comes out of a discharge line.

36. The water-tank-to-pump test is important because it tests the piping and valves between the water tank and the fire pump.

37. No, the piping is too small. The exception to this is an ARFF truck, whose pump must be able to obtain full flow from the water tank.

38. Mechanical seal

39. Stuffing box seal

40. The lantern seals may be displaced, the water flow stopped, and the shaft scored due to overheating.

41. 8–120 drops/minute with between 100 psi (689 kPa) and 150 psi (1034 kPa) in the pump system.

42. The pump and the engine must be shut off to prevent the maintenance worker from being caught in the driveshaft.

43. The Darley fire pump with injection-type packing may be adjusted from outside the truck body while the pump and engine are running.

Chapter 8

1. Governor pressure, modulator pressure, shift lever location

2. Yes

3. A deep ratio transmission has a very high-ratio first gear; this helps develop very high torque. A close ratio transmission has gear ratios that are closely spaced. Both transmissions will have the same ratios in their highest gear.

4. The HT 700 uses a screen in the oil pan and an external paper pleated filter usually found on the truck frame.

5. No, cork gaskets are not recommended.

6. The SAM plate gives the technician valuable information on the transmission's serial number, model number, and assembly number.

7. The SAM plate is located at the right-hand side of the rear of the transmission.

8. To send information to the transmission's valve body regarding the rotational speed of the output shaft of the transmission

9. The governor is exposed to transmission main hydraulic pressure. As the transmission output shaft and the governor rotate, centrifugal weights open passages in the governor. These passages are connected to the transmission's valve body. The faster the governor rotates, the higher the pressure the valve body receives.

10. Inked numbers on the governors' heads

11. True

12. High vortex flow occurs during hard acceleration when the torque converter is multiplying engine torque. This condition can occur when pulling heavy loads or driving up long, steep gradients.

13. Yes, it does.

14. Torque converters multiply only torque.

15. Yes, both the engine's torque curve and the torque converter ratio were initially selected based on tire size and rear axle ratio; any changes can affect vehicle performance.

16. Check the SAM number and refer to the parts manual.

17. The transmission's oil cooler removes the heat. Transmissions with retarders will require larger oil coolers than transmissions without.

18. The oil will degrade; varnish will form on the valves and cause erratic operation. This in turn could mean clutch packs do not receive the correct oil pressure and flows causing them to slip. Slipping clutch plates and discs will be destroyed (burned and warped) very quickly by friction.

19. It proves that the oil cooler is functioning and the torque converter's stator is freewheeling.

20. Most likely the torque converter's stator is defective.

21. No, as long as the transmission oil does not overheat.

22. The transmission must be in direct drive and the torque converter must be locked.

23. They are accomplished by external electrical valving and plumbing. Internal modifications may also be necessary.

24. Through the use of inputs and outputs to the transmission

25. False

26. Fourth gear

27. Because this information is also available with a diagnostic data reader connected to the transmission's CPU

28. No

29. Five

30. An adaptive learning program in the CPU

31. Speed sensors are inputs.

Glossary/Index

Glossary

A

Acceptance Tests — Preservice tests performed at the factory or after delivery to assure the purchaser that the apparatus meets bid specifications.

**Aerial Device — An aerial ladder, elevating platform, aerial ladder platform, or water tower designed to position personnel, handle materials, provide continuous egress, or discharge water.

Aircraft Fire Apparatus — Fire apparatus specifically designed for aircraft crash fire-fighting/rescue operations.

ALCL — Assembly line communications link.

Alternator — An electron pump.

American Wire Gauge (AWG) — A system of sizing wire by diameter; the number is inversely proportional to the diameter of the wire (i.e., as the AWG number gets bigger the wire size gets smaller).

Ammeter — A device to measure electron flow.

Analog — A meter that indicates electrical values by the use of a needle or pointer.

Anode — The positive electrode, also used in fire pumps or water tanks to combat corrosion.

Antimony — A metal used in maintenance-free battery grids to add strength and reduce gassing.

Appliance — Generic term applied to any nozzle, wye, siamese, deluge monitor, or other piece of hardware used with fire hose for the purpose of delivering water.

**Approved — Acceptable to the "authority having jurisdiction."

Attack Lines — Hose lines or fire streams used to attack, contain, or prevent the spread of a fire.

Attack Pumper — A pumper that is positioned at the fire scene and is directly supplying attack lines.

**Authority Having Jurisdiction — The organization, office or individual responsible for approving equipment, an installation, or a procedure.

AWG - See *American Wire Gauge.*

B

Baro Sensor — A sensor to measure barometric air pressure.

Battery — A device that stores chemical energy and converts it to electrical energy.

Bleeder Valve — A valve on a gate intake that allows air from an incoming supply line to be bled off before allowing the water into the pump.

Booster Apparatus — See *Brush Apparatus.*

Booster Tank — See *Water Tank.*

*Bourdon Tube — A pressure gauge that has a curved, flat tube that changes its curvature as pressure changes; this movement is then transferred mechanically to a pointer on the dial.

**Breathing Air System — The complete assembly of equipment such as compressors, purification system, pressure regulators, safety devices, manifolds, air tanks or receivers, and interconnected piping required to deliver breathing air.

Brush Apparatus (Booster Apparatus) — A fire department apparatus designed specifically for fighting ground cover fires.

Brushes — Carbon or copper conductors that conduct electrical current from nonrotational components to rotational components.

C

Catch Basin — See *Portable Tank.*

Cathode — A negative electrode.

*Cavitation — A condition in which vacuum pockets form in a pump and cause vibrations, loss of efficiency, and possible damage.

CCA — See *Cold Cranking Amperes.*

Cell — A group of negative and positive plates that will produce 2.1 volts when exposed to an acid solution.

**Fire Terms,* Copyright © 1980, National Fire Protection Association, Quincy, MA 02269. Reprinted with permission.

**NFPA 1071, *Standard for Emergency Vehicle Technician Professional Qualifications,* Copyright © 2000, National Fire Protection Association, Quincy, MA 02269. Reprinted with permission.

*Centrifugal Pump — A pump with one or more impellers that utilizes centrifugal force to move the water.

Certification Tests — Preservice pump tests conducted by an independent testing laboratory before delivery of an apparatus. These tests ensure that the apparatus will perform as expected after being placed into service.

Charging Circuit — An electrical system whose components include an alternator, batteries, voltage regulator, and connecting wires.

Chassis — The frame upon which the body of the fire apparatus rests.

*Circulator Valve — A device in a pump that routes water from the pump to the supply to keep the pump cool when hose lines are shut down.

*Clapper Valves — A hinged valve that permits the flow of water in one direction.

Cold Cranking Amperes — A battery's rating based on the amount of amperage the battery can produce at 0° F (-18° C) for 30 seconds and still maintain a voltage of at least 1.2 volts per cell.

Compound Gauge — A gauge connected to the intake side of the pump that is capable of measuring positive or negative intake pressures.

Conductor — A material that conducts either electricity or heat.

D

Dead-End Main — A water main that is not looped and in which water can flow in only one direction.

**Defect — A condition in which a device's performance is not up to established standards or there is an identified fault.

**Deficiency — A condition in which the application of a component is not within its designed limits or specifications.

**Deformation — Abnormal wear, defects, cracks or fractures, warpage, and deviations from a device's original condition that would affect its safe and correct operation.

**Diagnosis — The determination of a problem's cause.

Digital — A signal characterized by an on/off waveform.

Digital Display — A method of display that uses numbers to display information instead of a pointer or a needle.

Discharge Velocity — The rate at which water travels from an orifice.

**Documentation — The process of gathering, classifying, and storing information.

Drafting Operation — The process of drawing water from a static source into a pump that is above the level of the water supply.

Drain Valve — A valve on pump discharges that facilitates the removal of pressure from a hoseline after the discharge has been closed.

Dual Pumping (also incorrectly called Tandem Pumping) — An operation in which a strong hydrant is used to supply two pumpers by connecting the pumpers intake-to-intake. The second pumper receives the excess water not being pumped by the first pumper, which is directly connected to the water supply source.

**Duty — A major subdivision of work performed by one individual.

E

Electricity — The flow of electrons from one atom to another.

Electromagnetic Induction — The induction of current from one wire to another. This does not require the wires to be touching each other or to be electrically connected.

Elevating Water Device — An articulating or telescoping aerial device added to some fire department pumpers to enable the unit to deploy elevated master stream devices. These devices range from 30 to 75 feet (9 m to 23 m) in height.

Elevation Loss — See *Elevation Pressure.*

Elevation Pressure — The gain or loss of pressure in a hose line due to a change in elevation.

Emergency Response Vehicle — Any motorized vehicle so designated by an organization or agency to respond to emergency incidents where provisions have been made to include warning systems and specialized components such as pumps, aerial devices, rescue equipment and is capable of transporting emergency response personnel.

Emergency Vehicle Technician (EVT) — An individual who performs inspection, diagnosis, maintenance, repair, and testing activities on emergency response vehicles and who by possession of a recognized certificate, professional standing, or skill, has acquired the knowledge, training, and experience, and has demonstrated the ability to deal with issues related to the subject matter, the work, or the project.

EMI — Electrical magnetic interference. See *RFI*.

EPROM — Erasable programmable read only memory. Sometimes referred to as EEPROM for electrically erasable programmable read only memory.

Equal — In terms of specifying apparatus, the same level of quality, standard, performance, or design, but not necessarily identical.

Equipment — In this manual, used to denote portable tools or appliances carried on fire apparatus.

External Water Supply — Any water supply to a fire pump from a source other than the vehicle's own water tank.

F

Fire Apparatus — Any fire department emergency vehicle that participates in fire suppression or other emergency situations.

Fire Apparatus Driver/Operator — A firefighter (Level II) charged with the responsibility of operating fire apparatus to, during, and from a fire or other emergency scene. The driver/operator is also responsible for routine maintenance of the apparatus and any equipment carried on it.

Fire Boat — A boat that carries large fire pumps and is capable of supplying master streams or supply line hose lines to land-based fire-fighting apparatus.

Fire Department — An organization or agency providing rescue, fire suppression, and other related activities. For the purposes of this [NFPA 1071] standard, the term "fire department" shall include any public, private, or military organization or agency engaging in this type of activity.

Fire Department Connection (FDC) — The point at which the fire department can connect into a sprinkler or standpipe system to boost the water flow in the system. This connection consists of a clappered siamese with two or more 2½-inch (65 mm) intakes or one large diameter (4-inch [100 mm] or larger) intake.

Fire Department Pumper — Fire apparatus having a permanently mounted fire pump with a rated discharge capacity of 500 gpm (2,000 L/min) or greater. This apparatus may also carry water, hose, and other portable equipment.

Fire Department Sprinkler Connection — See *Fire Department Connection*.

Fire Department Standpipe Connection — See *Fire Department Connection*.

Fire Hydrant — An upright metal casting that is connected to a water supply system and is equipped with one or more valved outlets to which a hose line or pumper may be connected to supply water for fire-fighting operations.

Fire Hydraulics — The science that deals with water in motion as it applies to fire-fighting operations.

Fire Stream — An unbroken stream of water applied to a fire for the purpose of controlling and extinguishing it.

Floating Dock Strainer — A strainer used for drafting operations that is designed to float on top of the water. This eliminates the problem of drawing debris into the pump and reduces the required depth of water needed for drafting.

Flowmeter — Mechanical device installed in a discharge line that senses the amount of water flowing and provides a readout in units of gallons (liters) per minute.

Fluid — A substance with no established shape, but capable of flowing from place to place. Fluid will

---

*Fire Terms, Copyright © 1980, National Fire Protection Association, Quincy, MA 02269. Reprinted with permission.

**NFPA 1071, *Standard for Emergency Vehicle Technician Professional Qualifications*, Copyright © 2000, National Fire Protection Association, Quincy, MA 02269. Reprinted with permission.

conform to the shape of the vessel in which it is contained. Fluids have very little adhesive quality and cannot be pulled or lifted, but must be pushed by having pressure applied and being confined in some manner to direct the flow. Fluids are divided into liquids or gases.

Fold-A-Tank — See *Portable Tank*.

Friction Loss — Loss of pressure created by the turbulence of water moving against the interior walls of a hose or pipe.

Full Fielding — Supplying full battery voltage and current to the field current (rotor) of an alternator.

Fuse — A safety device that will melt if it conducts excessive electrical current. Rated in amperage.

Fusible Link — A safety device that will melt if it conducts excessive electrical current.

G

Gassing — The release of oxygen and hydrogen from the plates of a battery. Can happen during recharge or discharge. *Warning:* hydrogen is explosive and oxygen aids in combustion.

*Gate — A control valve for hose, a pump outlet, or a large caliber nozzle.

Gradability — The ability of a piece of apparatus to traverse various types of terrain.

Ground — A current path that completes an electrical circuit back to the battery. In modern North American vehicles this is of a negative potential.

H

Hard Suction Hose — A flexible, rubber length of hose reinforced with a steel core to prevent it from collapsing. This type of hose is connected between a fire pump and a water supply source (usually a static supply) and must be used when drafting.

Head — Water pressure due to elevation. For every 1 foot increase in elevation, 0.434 psi are gained. (For every 1 meter increase in elevation, 10 kPa are gained.)

Head Pressure — See *Head*.

Horsepower — A unit of power: 33,000 foot-pounds per minute equals one horsepower; one horsepower equals 746 watts.

Hose Layout, Complicated — Hose layout that includes the use of multiple lengths of unequal hose lines, unequal wyed or manifold lines, siamesed, or master stream devices. Such a layout requires the pump operator to perform complicated calculations in order to supply the lines properly.

Hose Layout, Simple — Hose layout that includes the use of single hose lines or multiple, wyed, siamesed, or manifold lines of equal length.

Hydraulic Calculations — The process of using mathematics to solve a problem involving fire hydraulics.

I

*Impeller — The vaned, circulating member of the centrifugal pump that transmits motion to the water.

*Impeller Eye — The intake orifice at the center of a centrifugal pump impeller.

**Incident Management System (IMS) — An organized system of roles, responsibilities, and standard operating procedures used to manage and direct emergency operations.

Increaser — An adapter used to attach a larger hose line to a smaller one. The increaser has female threads on the smaller side and male threads on the larger side.

**Inspect — To determine through examination by sight, sound, or feel, the condition and operation of systems or components by comparing physical, mechanical, and/or electrical characteristics with established standards, recommendations and requirements.

Insulator — A material that does not readily conduct electrical current or heat.

Intake Hose (Hard Suction Hose) — A flexible, rubberized length of hose with a steel core that connects a pump to a source of water. Intake hose is most commonly used for drafting operations.

---

*Fire Terms, Copyright © 1980, National Fire Protection Association, Quincy, MA 02269. Reprinted with permission.

**NFPA 1071, *Standard for Emergency Vehicle Technician Professional Qualifications,* Copyright © 2000, National Fire Protection Association, Quincy, MA 02269. Reprinted with permission.

J

Job — The combination of duties a worker performs.

Job Performance Requirement (JPR) — A statement that describes a specific job task, lists the items necessary to complete the task, and defines measurable or observable outcomes and evaluation areas for the specific task.

Jumper Cables — Heavy-gauge electrical cables used to recharge a discharged battery from a serviceable battery. Also called booster cables.

K

Kilo — Prefix meaning 1000; often abbreviated K.

Kilopascal (kPa) — Metric unit of measure for pressure; 1 psi = 6.895 kPa; 1 kPa = .1450 psi.

Kink — Severe bend in a hose line that increases friction loss and reduces the flow of water through that hose.

L

L Ten Life — The point at which 10 percent of a group of bearings has failed (measured in hours or revolutions).

Labeled — Equipment or materials to which has been attached a label, symbol, or other identifying mark of an organization acceptable to the authority having jurisdiction and concerned with product evaluation, that maintains periodic inspection of production of labeled equipment or materials and by whose labeling the manufacturer indicates compliance with appropriate standards or performance in a specified manner.

LCD — Liquid crystal display.

Lead Peroxide — The material on the positive plate of a battery.

Lead Sulfate — The material that forms on both the positive and negative plates of a discharged battery.

LED — Light emitting diode.

Lift, Dependable — The height a column of water may be lifted in sufficient quantity to provide a reliable fire flow. Lift may be raised through a hard suction hose to a pump, taking into consideration the atmospheric pressure and friction loss within the hard suction hose. Dependable lift is usually considered to be 14.7 feet (4.48 m).

Lift, Maximum — The maximum height to which any amount of water may be raised through a hard suction hose to a pump.

Lift, Theoretical — The theoretical, scientific height that a column of water may be lifted by atmospheric pressure in a true vacuum. At sea level, this height is 33.8 feet (10 m). The height will decrease as elevation increases.

Light Attack Vehicle — See Minipumper.

Line Voltage Circuit, Equipment, or System — An AC or DC electrical circuit, equipment, or system in which the voltage to ground or from line to line is 30 volts rms (AC) or 42.4 volts peak (DC) or greater, but does not exceed 250 volts rms (AC) or peak (DC).

Listed — Equipment or materials included in a list published by an organization acceptable to the "authority having jurisdiction" and concerned with product evaluation, that maintains periodic inspection of production of listed equipment or materials and whose listing states either that the equipment or material meets appropriate standards or has been tested and found suitable for use in a specified manner.

Loading Site — The point in a tanker shuttle operation where apparatus tanks are filled from a water supply.

Low Voltage Circuit, Equipment, or System — An electrical circuit, equipment, or system where the voltage does not exceed 30 volts rms (AC) or 42.4 volts peak (DC), usually 12 volts DC in fire apparatus.

Lugging — A condition that exists when the engine is operating at full throttle but below rated speed. Lugging can be eliminated by using a lower gear and proper shifting techniques.

Fire Terms, Copyright © 1980, National Fire Protection Association, Quincy, MA 02269. Reprinted with permission.

**NFPA 1071, *Standard for Emergency Vehicle Technician Professional Qualifications,* Copyright © 2000, National Fire Protection Association, Quincy, MA 02269. Reprinted with permission.

M

Maintenance — The act of servicing an emergency response vehicle or a component within a scheduled time frame prescribed by the manufacturer or department SOP to keep the vehicle and its components in operating condition. Maintenance includes minor repair such as replacing missing fasteners, changing lamps, fuses, and belts; and removing and replacing small self-contained components.

Manifold — (a) A hose appliance that divides one larger hose line into three or more smaller hose lines. (b) A hose appliance that combines three or more smaller hose lines into one large hose line. (c) The top portion of a pump casing.

Manufacturers' Specifications — Any requirement or service bulletin an emergency response vehicle builder or component producer provides with regard to the use, care, and maintenance of its product(s).

Manufacturer's Tests — Fire pump tests performed by the manufacturer prior to delivery of the apparatus.

MAP — Manifold absolute pressure.

Midipumper — An apparatus sized between a mini-pumper and a full-sized fire department pumper, usually with a gross vehicle weight of 12,000 pounds (5,443 kg) or greater. The midipumper has a fire pump with a rated capacity generally not greater than 1,000 gpm (4,000 L/min) .

Midship Pump — A fire pump mounted on the apparatus frame between the front and rear axle.

Minipumper — A small fire apparatus mounted on a pickup-truck-sized chassis, usually with a pump having a rated capacity less than 500 gpm (2,000 L/min). The minipumper's primary advantage is speed and mobility, which enables it to respond to fires more rapidly than larger apparatus.

Mobile Water Supply Apparatus (Tanker, Tender) — A fire apparatus with a water tank of 1,000 gallons (4,000 L) or larger whose primary purpose is transporting water. The truck may also carry a pump, some hose, and other equipment.

Multistage Pump — Any centrifugal fire pump having more than one impeller.

Mutual Induction — The generation of an electrical current in a coil of wire by a magnetic field.

N

Net Pump Discharge Pressure (NPDP) — The actual amount of pressure being produced by a pump. When taking water from a hydrant, NPDP is the difference between the intake pressure and the discharge pressure. When drafting, it is the sum of the intake pressure and the discharge pressure. (NOTE: Intake pressure is credited for lift and intake hose friction loss and is added to the discharge pressure.)

Nonconforming Apparatus — An apparatus that does not conform to the standards set forth by NFPA 1901, Standard for Automotive Fire Apparatus.

Nozzle Pressure — The velocity pressure at which water is being discharged from a nozzle.

Nozzle Reaction — The counterforce from the velocity of water being discharged, directed against the people or device holding the nozzle.

O

Ohm — A unit of electrical resistance. One amp of current will flow through a wire with one ohm of resistance if there is a potential of one volt across the wire.

Ohmmeter — A test instrument used to measure electrical resistance.

Open Circuit — The condition that exits whenever there is no electron flow.

Orifice — An opening through which a substance may pass.

Oscilloscope — A visual display of the electrical current on a cathode ray screen.

P

Parallel Operation — See *Volume Operation*.

Performance Requirements — A written list of expected capabilities for new apparatus that is produced by the purchaser and presented to the manufacturer as a guide.

Fire Terms, Copyright © 1980, National Fire Protection Association, Quincy, MA 02269. Reprinted with permission.

**NFPA 1071, *Standard for Emergency Vehicle Technician Professional Qualifications,* Copyright © 2000, National Fire Protection Association, Quincy, MA 02269. Reprinted with permission.

Piezoelectric — An electrical current that is generated within certain crystals when pressure is applied to them.

Portable Tank (also Port-A-Tank, Fold-A-Tank, Portable Basin) — A collapsible storage tank used during relay or shuttle operations to hold water from water tanks or hydrants. This water can then be used for supplying attack apparatus.

***Positive Displacement Pump** — A pump that moves a given amount of water through the pump chamber with each stroke or rotation. Positive displacement pumps are self-priming.

Power Plant — An apparatus's engine.

***Power Take-Off (PTO)** — A device used to transmit power through a clutch arrangement from the power plant to the auxiliary equipment.

Preservice Tests — Tests performed on fire pumps before they are put into service. Preservice tests are broken down into manufacturer's tests, certification tests, and acceptance tests.

Pressure Gauge — A gauge that registers the pump discharge pressure.

Pressure Governor — A pressure control device designed to eliminate hazardous conditions resulting from excessive pressures by controlling engine speed.

Pressure Operation (Series) — The operation of a two-stage or more centrifugal pump in which water passes consecutively through each impeller to provide high pressures at a reduced volume.

Prime — To remove all air from a pump in preparation for receiving water under pressure.

Priming Pump (Primer) — A small positive displacement pump used to evacuate air from a centrifugal pump housing and hard suction hose. Evacuating air allows the centrifugal pump to receive water from a static water supply source.

PROM — Programmable read only memory.

PTO — See *Power Take-Off.*

Pulse Width Modulation — The electrical control of a solenoid or valve by rapidly turning on or off the electrical current to that solenoid or valve. Measured in a percentage of on or off time.

Pump and Roll — The ability of an apparatus to pump water while the vehicle is in motion.

Pump Capacity Rating — The maximum amount of water a pump will deliver at the indicated pressure.

Pump Charts — Charts carried on a fire apparatus to aid the pump operator in determining the proper pump discharge pressure to use when supplying hose lines.

Pump Discharge Pressure — The actual velocity pressure (measured in pounds per square inch) of the water as it leaves the pump and enters the hose line.

Pump Drain — A drain located at the lowest part of the pump to help remove all water from the pump. This is done to eliminate the danger of damage due to freezing.

Pump Operator — A firefighter charged with operating a pump and determining the pressures required to operate it efficiently.

Pumping Apparatus — A fire department apparatus whose primary responsibility is to pump water.

Q

****Qualified Person** — A person who, by possession of a recognized degree, certificate, professional standing, or skill through knowledge, training, and experience, has demonstrated the ability to deal with problems related to the subject matter, the work, or the project.

R

Radio Choke — A device used to reduce RFI or EMI; also called a torrid coil.

****Rebuild** — To make extensive repairs, to restore a component to like new condition in accordance with original manufacturer's specifications.

Rectifier — A diode that converts AC current to DC current.

---

*Fire Terms, Copyright © 1980, National Fire Protection Association, Quincy, MA 02269. Reprinted with permission.

**NFPA 1071, *Standard for Emergency Vehicle Technician Professional Qualifications,* Copyright © 2000, National Fire Protection Association, Quincy, MA 02269. Reprinted with permission.

Rectifier Bridge — Usually a group of six diodes, three negative and three positive. Used to rectify an alternator's three-phase alternating current to direct current.

Reducer — An adapter used to attach a smaller hose to a larger hose. The female end has the larger threads, while the male end has the smaller threads.

Relay — An electromagnetic device that uses a very small amount of current to control a large amount of current. The two general types of relays are normally open (NO) and normally closed (NC).

Relay Pumping — The process of using two or more pumpers to move water through hose lines over a long distance by operating the pumpers in series.

Relief Valve — Pressure control device designed to eliminate hazardous conditions resulting from excessive pressures by allowing this pressure to bypass to the intake side of the pump.

**Repair — To restore to sound condition after failure or damage.

**Requisite Knowledge — The necessary knowledge one must have in order to perform the assigned task.

**Requisite Skills — The necessary skills one must have in order to perform the assigned task.

Reserve capacity — The number of minutes a battery can continue to deliver 25 amps with a voltage no lower than 1.75 volts per cell.

Residual Magnetism — The magnetic force remaining in an object after an electromagnetic force has been removed from that object. Often used in connection with large brushless alternators.

*Residual Pressure — The pressure remaining in a system while water is flowing.

Response District — The geographical area where a particular apparatus has a first-due assignment for a fire or other emergency incident.

RFI — Radio frequency interference; can disrupt sensitive sensors or electrical components. Also called EMI.

ROM — Read only memory.

RPM — Revolutions per minute.

S

SAE —The Society of Automotive Engineers. Coupled with a number, these initials indicate motor oil viscosity.

Safety Bar — A hinged or sliding bar designed to protect firefighters from falling out of the jump seat area of a fire apparatus.

Sediment — Dirt and other foreign debris carried along in water; sediment may collect in water-moving equipment.

Sediment Chamber — A container in a fuel system where contamination settles out of the fuel.

Series Operation — See *Pressure Operation.*

Service Tests — A series of tests performed on a fire pump to ensure its operational readiness. Service tests should be performed at least yearly or whenever apparatus has undergone extensive repair.

Servo Unit — A device that uses an electrical or electronic input signal to control a mechanical component.

**Shall — Indicates a mandatory requirement.

Short-Circuit — An unintentional copper-wire-to-copper-wire connection.

Short to Ground — An unintentional copper-wire-to-steel-frame connection.

**Should — Indicates a recommendation or that which is advised but not required.

*Siamese — A hose appliance used to combine two or more hose lines into one. It generally has female inlets and a male outlet.

Single-Stage Centrifugal Pump — A centrifugal pump with only one impeller.

*Size-Up — The mental evaluation made by a firefighter or officer that enables him or her to determine a course of action.

Soft Sleeve Hose (also called Soft Suction) — A large diameter, collapsible piece of hose used to connect a fire pump to a pressurized water supply source.

Solenoid — An electromagnetic device that produces mechanical motion. Used to control hydraulic oil.

Specifications — (a) Detailed information provided by a manufacturer on the function, care, and maintenance of equipment or an apparatus. (b) A detailed list of requirements prepared by a purchaser and presented to a manufacturer or distributor when purchasing equipment or apparatus.

Speedometer — A dashboard gauge that measures the speed at which the vehicle is traveling. It may also indicate proper operation of the fire pump.

***Sprinkler Head** — The water flow device in a sprinkler system. The sprinkler head consists of a threaded nipple that connects to the water pipe, a discharge orifice, a heat-actuated plug that drops out when a certain temperature is reached, and a deflector that creates a stream pattern suitable for fire control.

Sprinkler System — A system of water pipes and sprinklers installed in a structure to automatically control and/or extinguish fires.

Staging — The process by which noncommitted units responding to a fire or other emergency incident report to a location away from the fire scene to receive their assignment.

Standard Apparatus — Any apparatus that conforms to the standards set forth by NFPA 1901, *Standard for Automotive Fire Apparatus.*

****Standard Operating Procedures** — Written guidelines or procedures created by the AHJ (authority having jurisdiction) or their designee for internal operation of the department or agency.

Standpipe System — A wet or dry system of pipes and fire hose outlets in a large single-story or multistory building. The standpipe system is used to provide for quick deployment of hose line during fire-fighting operations.

State of Charge — The amount of battery charge. On sealed maintenance-free batteries, this is measured with a digital voltmeter.

Static Pressure — The pressure exerted in all directions at a point in a fluid at rest.

Static Water Source — A nonpressurized source of water from which pumpers may draft for fire-fighting or other operations.

Steamer Connection — The 4½-inch (115 mm) or larger connection on a fire hydrant.

Stepper Motor — An electrical motor that moves a very precise distance or number of revolutions when a given amount of electrical current is applied.

****Structural Integrity** — An unimpaired condition of any component. (from NFPA 1071 Standard)

T

Tachometer — A dashboard or pump-panel gauge that measures engine speed in revolutions per minute.

Tandem Pumping — A short relay operation in which the pumper taking water from the supply source pumps into the intake of the second pumper. The second pumper boosts the pressure of the water even higher. Tandem pumping is used when pressures higher than the capability of a single pump are required.

Tanker — See *Mobile Water Supply Apparatus.* Also, in the ICS System, *tanker* refers to a water transporting fixed wing aircraft.

****Task** — An essential step of a work operation required to complete the performance of a duty.

Tender — See *Mobile Water Supply Apparatus.*

****Test** — To verify serviceability by measuring the mechanical, pneumatic, hydraulic, or electrical characteristics of an item and comparing those characteristics with prescribed standards.

Thermistor — A resistor that changes resistance with temperature. Can be used to measure accurately engine coolant temperature or oil temperature. A negative coefficient increases resistance with a decrease in temperature. A positive coefficient increases resistance as temperature increases.

Fire Terms, Copyright © 1980, National Fire Protection Association, Quincy, MA 02269. Reprinted with permission.

**NFPA 1071, *Standard for Emergency Vehicle Technician Professional Qualifications,* Copyright © 2000, National Fire Protection Association, Quincy, MA 02269. Reprinted with permission.

Torque — A twisting force.

Total Pressure — The total amount of pressure loss in a hose assembly due to friction loss in the hose and appliances, elevation losses, or any other factors.

Traffic Control Device — A mechanical device that automatically changes traffic signal lights to favor the path of responding emergency apparatus.

Transducer — A device that converts speed, pressure, or distance into electrical signals.

Transistor — A semiconductor that can be used as an electrical relay switch or amplifier.

Triple-Combination Pumper — A fire department pumper that carries a fire pump, hose, and a water tank.

Two-Stage Centrifugal Pump A centrifugal pump with two impellers.

U

Unloading Site — The point in the tanker shuttle operation where portable tanks are located and the tankers unload their water.

V

Vacuum — A pressure below atmospheric pressure; measured in inches of mercury (Hg) or kilopascals (kPa).

Volt — A measure of electrical pressure.

Voltmeter — A test meter used to measure electrical pressure.

Volume Operation — (Parallel) The operation of a two-stage or more centrifugal pump in which each impeller discharges into a common outlet, thereby providing the maximum flow at the rated pressure.

*****Volute** — The spiral, divergent chamber of a centrifugal pump in which the velocity energy given to water by the impeller blades is converted to pressure.

W

Warning Device — Any audible or visual device added to an emergency vehicle to gain the attention of drivers of other vehicles. Warning devices may include flashing lights, sirens, horns, or bells.

Water Distribution System — A system designed to supply water for residential, commercial, industrial, and/or fire protection purposes. A water distribution system comprises a network of piping and pressure-developing equipment.

Water Supply — Any source of water available for use in fire-fighting operations.

Water Supply Pumper — A pumper that takes water from a source and sends it to attack pumpers operating at the fire scene.

Water Tank (Booster Tank) — A water storage receptacle carried directly on the apparatus. NFPA 1901 specifies that Class A pumpers must carry at least 300 gallons (1,200 L) of water and minipumpers must carry at least 150 gallons (600 L) of water.

Watt — A measure of electrical power.

Wye — A hose appliance that divides one hose line into two hose lines of equal or smaller size.

*Fire Terms, Copyright © 1980, National Fire Protection Association, Quincy, MA 02269. Reprinted with permission.

**NFPA 1071, Standard for Emergency Vehicle Technician Professional Qualifications, Copyright © 2000, National Fire Protection Association, Quincy, MA 02269. Reprinted with permission.

Index

braking, engine, and modulated lockup 244

breather, transmission 252

breathing air compressor, and portable-engine-driven alternators 185

breathing protection, repair shop 23

brush/booster apparatus 41

burn-in time, electronic devices 213

C

calcium, engine location 152

Canada Motor Vehicle Safety Standards (CMVSS) 12

carboxylate (COX), in fluid analysis 154

cavitation

 and electronic pump governor 212

 and fluid analysis 145-147

 and fluid velocity 205

 overview 72-74

cell phones, and RFI 195

central processing unit (CPU), electronic pump governor 212

certification tests, pump. *See also specific test types* 217-220, 222-224

changeover valve. *See* volume-pressure transfer valve

chassis maintenance 102, 112-113

chromium, engine location 152

cleaning, apparatus 91-92, 102

cleaning, parts, safety practices 24-27

cleaning solvents, safety practices 26-27

close-ratio (CR) transmission 240

clothing, personal protective, for EVT 11-12

clothing standards, and workplace safety 22-23

clutch, one-way over-running, torque converter 247

clutch packs, and stall test 248

coaxial shield cable, and RFI 196

cold cranking amps (CCA) 108

commercial garages, evaluating and using 6

compressed air, safety practices 19

computers, power for 180-186

contamination control number, hydraulic oil 140

coolant, engine

 and cavitation 145

 electric currents in 146, 161-166

 maintenance 97

coolant heaters, and coolant overheating 146

cool-down test, automatic transmission 247

cooler, air-to-air, and stall test results 248

cooling system maintenance 161-166

cooling system, auxiliary

 and engine speed control 214

 maintenance 101

copper, engine location 152

corrosion, and electrical problems 161

crane, hydraulic floor 20

crisis maintenance 13-14

cross-mounted engine drive 48-49

cutting wear, and ferrography 154

D

dangerous goods. *See* hazardous materials

decontamination, apparatus 11-12

deep-reduction transmission 240

diesel engines

 electronically controlled 210

 mechanically controlled 210

diesel engines, two-cycle

and multigrade oils 130-131

differential oil 100

differentials, rear-end, and electrical current in coolant 164

diodes

 as cause of electrical problems 179

 cooling medium 177

 heat damage 177-178

 heat sinks, cleaning 176

 location on apparatus 177

 testing procedure 179-180

direct range, transmission, during pumping operations 249

DN factor 135-136

do-not-shift (DNS) condition, in electronic transmissions 254

door maintenance 96

drain maintenance 101

drive shaft

 balance weights 113

 maintenance 102

drive system, and pumping problems 225

driveline component maintenance 112-114

driver/operators

 and apparatus safety 9

 licensing requirements 12-13

 role in apparatus maintenance 8

driving area maintenance 92-94

D-S-O (drive-supply-output) sequence 224-226

E

electrical connections

 and electronic pump governor 213

 faulty, and ECU 255

 for fire-pump transmission modifications 251

 steps in making 109-110

 troubleshooting Allison WT transmissions 259

electrical current in coolant 146, 161-166

electrical power, 110-volt AC

 sine wave 183

 square wave 186

 supplying 180-186

electrical power, DC, and RFI 193-196

electrical problems, rectifying 179-180

electrical test for voltage in cooling system

frequency devices. *See* electronic devices

frequency meter, hydraulically driven alternator 183

friction, and heat analysis 159

friction and entrance loss (FEL)

    calculating 205

    and net pump pressure 205

friction loss, in piping and hose 209

front-mount pumps 50-51

fuel filter, plugged, and stall test results 248

fuel injector, defective, and stall test results 248

fuel level maintenance 98, 116

fuel pump maintenance 103

fuel sediment bowl maintenance 102

fuel tanks, NFPA requirements 116

fuel tank capacity, and pumping test 220

G

gauges, maintenance of 92

gear box, fire pump

    and heat analysis 159

    inspection 114-115

    and oil analysis 147

generator. *See* alternator

governor, automatic transmission 241-242

governor, fire pump 115

governor filter, transmission 242

grease 133-138

greasers, automatic, and grease grades 135

grinder, bench, safety practices 16-18

ground, electrical, and coolant problems 162, 164

H

hand light maintenance 104

hardness, grease 135

hazardous materials

    categories of 32

    disposal of 29-30

    safety practices 25-26

head gasket failure, and electrical problems 162

head pressure 61-62

headlight maintenance 104

hearing protection 12

heat analysis, and lubrication program 159

heater/defrost system maintenance 105

hoist, chain, safety practices 20

hose connection maintenance 101

hose load maintenance 104

hose-reel rewind maintenance 105

housekeeping safety practices 27-29

humidity, relative, and engine performance 202

hydrant, fire, color coding 204

hydrant pressure and cavitation 75

hydrostatic pump test requirements 215, 216

I

impeller, centrifugal pump 67

infrared heat gun, in heat analysis 159

initial attack fire apparatus 42. *See also* minipumper and midipumper

inputs, electronic transmission 254

intake valves, and stall test results 248

interlock, pump governor 213

International Association of Fire Chiefs (IAFC), and AMS, 3

International Standards Organization for Solid Contamination Control oil cleanliness ratings 139

inverter

    to power electronic devices 185-186

    and RFI 186

    wiring for 187-188

iron, engine location 152

ISO 4572 test filtration ratio 142

ISO oil cleanliness ratings

    and fluid analysis 139-143

    ISO 4572 test 142-143

    and spectrographic analysis 153

J

jack safety practices 19

K

kinematic viscosity. *See* viscosity

L

L 10 life 175-176

ladder apparatus. *See* aerial apparatus, ladder

ladders, portable, maintenance 104

lead, engine location 152

lift safety practices 19

lifting and carrying safety practices 23

lights

    maintenance 96, 104

    rotating 105

lithium, grease additive 136

lockup, high-range, in transmission fire-pump mode 249

long-range planning, importance of 6

lubricants, bearing 133-134

lubrication

    engine 112

    overview 129

    schedules 103

lubrication system, and engine speed control 214

M

magnesium, engine location 152

maintenance, cost of, and public trust 5

maintenance, general

    daily 92-99

    periodic 102-103

    weekly 99-102

maintenance equipment safety practices 22

maintenance levels 13-16

manganese, engine location 152

manufacturer's specifications vs. department SOPs 13

material safety data sheets (MSDS) 30-32

mechanical engines. *See* nonelectronic engines

midipumper 43

midship transfer drive 51

minipumper 42-43

mirror maintenance 92

mobile water supply fire apparatus 43-44

modulated lockup, torque converter 244

modulator valve, for automatic transmission 242

molybdenum

    engine location 152

    grease additive 136

motor carrier regulations 12

motors, electrical, maintenance 105

multiple-stage centrifugal pump 69, 71-72

N

National Association of Emergency Vehicle Technicians
(NAEVT) 4-5

National Lubricating Grease Institute (NLGI) grease grades
135

National Transportation Safety Board (NTSB) *Special Investigative Report, Emergency Fire Apparatus* 3

NFPA 1071, *Standard for Emergency Vehicle Technician Professional Qualifications*

    and department size 6

    scope and purpose 4

NFPA 1500, *Standard on Fire Department Occupational Safety and Health Program* 9-10

NFPA 1581, *Standard on Fire Department Infection Control Program* 11-12

NFPA 1904, *Standard on Aerial Ladder and Elevating Platform Fire Apparatus* 117

NFPA 1911, *Standards for Service Tests of Fire Pump Systems on Fire Apparatus* 211

NFPA 1915, *Standard for Fire Apparatus Preventive Maintenance Program* 4

nickel, engine location 152

nitrous oxide (NOX), in fluid analysis 154

nonelectronic engines, electronic pump controls for 213

normal abrasive wear, and ferrography 154

normal operating pressure 662

nozzle maintenance 104

O

Occupational Safety and Health Administration (OSHA), and
hazardous materials 31

oil. *See also specific apparatus component requirements*

    brands of, mixing 147-148

    color of 156

    container size 148

    degradation of, in oil cleanliness report 154

    vs. grease 133

    physical properties of, in oil cleanliness report 153

oil, extra pressure (EP) gear, for pump transfer case 147

oil, hydraulic

    cleanliness requirements 140-41

    for hydraulically driven alternator system 182

oil, motor

    grades 130-132

    ISO ratings 141

    in radiator coolant 146

oil additives, antirust 103

oil analysis report 149-159

oil change intervals, and fluid analysis 147

oil cleanliness ratings. *See* ISO oil cleanliness ratings

oil cooler

    and electrical current in coolant 166

    and electrical problems 161

oil cooler transmission 245

oil filter ratings

    absolute 142

    filtration ratio 142-143

    nominal 141-142

oil filters. *See also specific apparatus components*

    and fluid analysis 141, 156

    ISO 4572 test 142-143

    maintenance 103

    visual inspection of 143

oil-level sensor in WT transmission 256

organizational chart 7

oscilloscope, for checking alternator output 179-180

output, water, and pumping problems 225

outputs, electronic transmission 254

overload test procedure and requirements 220

oxidation (OXD), and fluid analysis 154, 156

P

parts washer safety practices 26-27

passenger safety, apparatus 10

penetration, property of grease 135

personal protective clothing. *See* clothing, personal protective

personal safety 24-32

phosphorous, engine location 152

pitting, and electrical problems 161, 163

polarizing alternators 174

polymers, in multigrade motor oil 130-132

portable extinguisher maintenance 104

portable pumping units 47

positive displacement pump. *See* pump, positive displacement

potassium, engine location 152

power source hour meter, for hydraulically driven alternator

Notes

Notes